Open Spaces, Open Rebellions

Open Spaces, Open Rebellions

The War over America's Public Lands

Michael J. Makley

University of Massachusetts Press
Amherst and Boston

Copyright © 2017 by University of Massachusetts Press
All rights reserved
Printed in the United States of America

ISBN 978-1-62534-314-7 (paper); 312-3 (cloth)

Designed by Sally Nichols
Set in Adobe Minion Pro
Printed and bound by Maple Press, Inc.

Cover design by Jack Harrison
Cover photo of Organ Pipe Cactus National Monument, U.S. by Irina K. Shutterstock.com.

Library of Congress Cataloging-in-Publication Data

Names: Makley, Michael J., author.
Title: Open spaces, open rebellions : the war over America's public lands / Michael J. Makley.
Description: Amherst : University of Massachusetts Press, [2017] | Includes bibliographical references and index.
Identifiers: LCCN 2017017183| ISBN 9781625343147 (pbk.) | ISBN 9781625343123 (hardcover)
Subjects: LCSH: Public lands—United States—History. | Land use—Government policy—United States—History.
Classification: LCC HD216 .M25 2017 | DDC 333.10973—dc23
LC record available at https://lccn.loc.gov/2017017183

British Library Cataloguing-in-Publication Data
A catalog record for this book is available from the British Library.

To Matt "Becks" Becker, Dan Makley, and
Matthew Makley for their invaluable assistance

Contents

Preface xi

Introduction	1
1. Creating the Federal Domain	9
2. Interior Battles	16
3. Rangeland Battles	22
4. Multiple Use	30
5. The Radical New Conservation	38
6. The Sagebrush Rebellion Begins	45
7. Harnessing Forces	55
8. Fracturing Policies	64
9. Inciting the Populace	74
10. County versus Federal Government	83
11. Extinguished Rights	88
12. Pursuing Ideology in the Courts	94
13. Pursuing Ideology with Guns	101
14. Opaque Governance	108
15. Differing Values	115
Conclusion	122
Notes	125
Index	147

Illustrations follow page 82

Preface

Before he moved from the West, my longtime editor and friend Becks Becker proposed a project based on the Sagebrush Rebellion of the 1980s. The rebellion was a movement in the thirteen farthest west states to transfer much of America's public land to the states or sell it into private ownership. I have lived my entire life in the rural West. In the 1970s I worked summers for the U.S. Forest Service at Lake Tahoe, and I have written books about the West, including an environmental and political history of Lake Tahoe. Owing to my experiences and interviews of individuals involved with the issues, I knew that the sentiments that drove the Sagebrush Rebellion still existed in the twenty-first century.

As I researched the history of federal land disputes, tracing them to the Theodore Roosevelt administration, a recursion of the rebellion flared, gaining national attention. A Nevada rancher, supported by armed militia, engaged in a standoff with federal agents over the rancher's refusal to pay for grazing his cattle on public lands. Similar incidents involving mining and off-road-vehicle (ORV) access followed, with armed protesters claiming the right to ignore public land laws and regulations. At the same time, significant amounts of money from development interests were invested in the movement to convey the lands from federal control, and elected representatives in several state legislatures and the U.S. Congress pursued transference legislation. This book explores the history that led to today's disputes and the ongoing issues that compose its volatility.

The gentlemen to whom this book is dedicated influenced and helped shape it. Over the course of many months, Becks Becker and I discussed the direction of the enterprise. Dan Makley provided large amounts of research that informed it. Matthew Makley, as on all my writing ventures, conferred with me and suggested solutions to problems, substantially improving the manuscript.

A number of other people also assisted in this project. I am grateful to Alicia Barber for providing access to the oral histories that factor into the chapters on Nevada and the Sagebrush Rebellion. Dominique and Jacques Etchegoyhen offered valuable information regarding land-conservation law and federal and state land programs. Larry Schmidt and Robert Stewart supplied material about the role of public agencies in specific land disputes. As is often the case, McAvoy Layne acted as a consultant, suggesting important avenues of research.

I wish to thank as well Mary V. Dougherty, director of the University of Massachusetts Press; production manager Carol Betsch, who shepherded the book through the publication process; Annette Wenda, who, as she has done for past projects, used her expertise as a copy editor to ensure readability and cohesion in the text; and Sally Nichols and Jack Harrison, who provided graphic planning and the book's design.

Additionally, the assistance of staffs at a number of libraries was critical to the story's development. These include the Western History Section of the Denver Public Library, University of Nevada Special Collections Department, Nevada State Library, Boise State University Library, Boise Public Library, California Section of the California State Library, Bancroft Library, and Institute of Governmental Studies Library at the University of California, Berkeley. Coi Drummond-Gehrig, with the Denver Public Library, was especially helpful, doing extra work in finding relevant photographs. Jessica Maddox, in the Nevada Special Collections Department, has consistently assisted me in finding important research, and she did so again. Sarah Alex, executive director of the Herb Block Foundation, provided the Herb Block cartoon, ensuring resolution that enabled its printing.

I am also immensely grateful to two reviewers without whom this book would be significantly less developed. Leisl Carr Childers suggested sources that provided breadth and depth to the work. She also convinced me to reorganize and rewrite parts of the manuscript, condensing historical chapters in favor of emphasizing recent history. I also followed the recommendations of Jodi Peterson, senior editor of *High Country News*.

She influenced me to clarify the constitutional arguments of the proponents of state claims to federal land and those of historical scholars. I also followed her suggestion to discuss the larger implications of transference on the nation as a whole.

Finally, thank you to Randi Makley, who is my continual source of support and inspiration.

Open Spaces, Open Rebellions

Introduction

Determined citizens, elected officials, and corporate interests are fueling a rebellion to remove public lands from federal jurisdiction. They contend that the central government is infringing on economic opportunities and states' rights. The lands, located almost entirely in the West, are a common inheritance of all Americans. Allowing their sale or closure for resource extraction must be weighed against people's right to access.

Efforts to gain state control or to privatize federal lands began when the government curtailed land sales and tightened regulatory policies at the end of the nineteenth century. The government previously had encouraged settlement or transferred lands to private corporations to promote the economy.

With the dawning realization that the country's forests were not unlimited and its grazing lands not indestructible, the government changed policy. At the time, conflicts over land use had become pronounced. Specific reasons for the change of direction included range wars, overgrazing of rangelands, the effects of clear-cutting forests, contaminated water downstream from grazing and logging operations, and the desire to set aside iconic lands.

Protestations by land users intensified dramatically early in the twentieth century when Theodore Roosevelt increased the amount of protected land from 42 million acres to 172 million. Fifty years later, objections escalated again as Congress determined that lands needed to be managed for multiple use. Along with availability for residences and agricultural and commercial use, the government set aside lands for the good of the environment and endangered species as well as for recreation.

Although present-day dissension has marked similarities to previous public land disputes, the current rebellion includes a violent edge rarely before in evidence. This violence came to the nation's attention in 2014 with the Cliven Bundy affair. It began twenty-one years earlier when the Bureau of Land Management (BLM) modified the Nevada rancher's grazing permit. In 1993, rather than follow the directive to limit animals in an area of environmentally fragile public lands, Bundy stopped paying grazing fees. In a show of contempt, he turned out nine hundred head of cattle in the section, almost nine times the number allowed by the new permit. Through the years, Bundy continued to use the federal range, though not acknowledging the government's authority over it. His unpaid fees and penalties rose to $1.1 million. In April 2014 the defiance erupted into open rebellion.[1]

After numerous attempts to get Bundy to pay or remove his herd from public lands, the BLM confiscated some of his cattle. They also notified him that, in enforcing two federal court orders, they intended to confiscate up to eight hundred head. The second time agents rounded up animals, they were confronted by hundreds of Bundy supporters and members of antigovernment militias, many brandishing automatic weapons. They trapped the officers in a desert wash. While the large group blocked their way, officers were fixed in the sights of snipers, including one who sprawled behind a concrete barrier on a highway overpass forty feet above. The militias' deployment was augmented by former Arizona County sheriff Richard Mack, who planned to use women as human shields. He said that he was willing to die for the cause, but "to show the world how ruthless these people are, women needed to be the first ones shot."

Bundy had rallied supporters from across the country by falsely claiming that Bureau of Land Management marksmen had surrounded his ranch, targeting family members. He sent out word that it was a "range war." On the Internet a Montana militiaman rallied others in what he called "Operation Mutual Aide," saying he wanted to "invite everyone to the first annual Patriots picnic at the Bundy Ranch." Before the conflict the militia groups gathered at a staging area, and Bundy asserted that "the federal government must get out of the State of Nevada." Behind him a Nevada State flag flew above an American flag. Confronting the federal agents in the wash, one of Bundy's sons, in his forties, declared, "You are on Nevada State property. You need to leave. . . . That's the terms." After a lengthy standoff, when a wrong move by either side might have sparked a shoot-out, the outmanned BLM agents retreated. The cattle they had gathered were released.[2]

In the midst of the episode, Fox News showed the sixty-eight-year-old Bundy in a white cowboy hat, on horseback, with an American flag billowing behind him. For days preceding the incident and afterward, television's Fox News host Sean Hannity lauded Bundy for challenging the national government. He was especially provocative when hosting Bundy and a son on air. He asserted that the government might want to make examples of the ranchers. And he asked Bundy's son if he feared federal agents coming late at night and perhaps killing members of the family.[3]

In Washington, DC, Democratic senator Harry Reid called the militiamen "domestic terrorists." A few Republican officials, though, voiced agreement with Hannity. U.S. senators Rand Paul and Ted Cruz termed the BLM actions government overreach. Nevada U.S. senator Dean Heller said the Bundy supporters were "patriots" and called on Congress to investigate the BLM's conduct.

Mark Potok, senior fellow at the Southern Poverty Law Center, had a different analysis, focusing on the Bundy defenders. "The Bundy ranch standoff wasn't a spontaneous response to Cliven Bundy's predicament," he said, "but rather a well-organized, military-type action that reflects the potential for violence from a much larger and more dangerous movement." Potok was referring to the battle plan worked out before the confrontation by paramilitary leaders. They had walked the site with Bundy days before, allowing riflemen to be placed so as to target the federal agents.[4]

The Bundy affair remained unresolved for two years. The BLM expressed disappointment that Bundy disregarded the laws that sixteen thousand public land ranchers abided by, and the agency maintained that it would pursue further legal action against him. But throughout 2015 the rancher, protected by armed bodyguards, continued to graze his cattle on public lands. At public rallies he spoke, claiming victory at having triumphed over the government. Finally, in 2016, after some of the principals were involved in other incidents of militias challenging federal authority, Bundy and those who abetted him were arrested. The key figures remain in jail, awaiting trial, at the publication of this book.

A year after the original Bundy incitement, in Washington, DC, an oft-recurring debate regarding public lands culminated with a telling vote. The Senate approved a nonbinding amendment to the budget bill, allowing states to acquire federal lands. The vote was fifty-one to forty-nine, with Republicans casting all fifty-one of the affirmative votes. The Republican Party has long espoused "responsible development" on federal lands. In the House of Representatives, the majority Republicans sponsored a number of bills authorizing the sale of public lands to private interests or the conveyance of lands to states.[5]

Bundy's insurrection and the Republican Party's congressional actions have amplified public land issues. This work examines the factors leading up to the intensification. It discusses the change in land policy away from resource development as its primary use and the resultant conflicts and rebellions.

Disputes arise in part because more than 90 percent of public lands, some 635–40 million acres, are in the thirteen states located farthest west. Ongoing contentions involve the region's ecology and quality of life weighed against economic growth.

Environmentalists cite the destruction and deterioration of lands and habitat as evidence of the need for regulation or preservation. They emphasize that the notion of the American frontier is sustained by untamed topographies, including vast deserts in the Southwest; forests—including remnants of primeval stands—in the Northwest; and innumerable mountain ranges and wildernesses. They observe that western states have benefited from their share of federal lands' licenses, royalties, leases, and fees.[6]

States' rightists, development interests, and political conservatives stress that concerns about production need to be given more weight, arguing against lands being kept from the marketplace. They insist the states would do better financially by administering the lands. They point to the imposition on the West due to the immensity of the national holdings. Percentages of public lands within state bounds include the following: Nevada, 81.1; Utah, 66.5; Alaska, 61.8; Idaho, 61.7; Oregon, 53.0; Wyoming, 48.2; California, 47.7; Arizona, 42.3; Colorado, 36.2; New Mexico, 34.7; Montana, 28.9; Washington, 28.5; and Hawaii, 20.3. Other contiguous states have minimal amounts of lands under federal control. Texas, for example, has only 1.8 percent.[7]

When regulatory agencies began managing forests and rangelands, they were dominated by the primary users: logging companies and stock growers' associations. Massive population growth in the second half of the twentieth century gave rise to the metropolitan West, creating a concurrent decline of political power of the rural West.[8]

In the 1960s the environmental movement, supported by Presidents Kennedy, Johnson, and Nixon, moved Congress to pass environmental acts, protecting air, water, wildlands, and endangered species. It was owing to these changes that multiple-use policies replaced the long-established preferences for extractive resources.

Agency administrators were charged with making difficult decisions about proper uses of land in specific areas. In rangelands, for example, attention had to be given first to the traditional use of grazing. Critical questions arose. What considerations should be allowed for generations of prior use? How can ranchers' improvements to the public range be compensated? Because a single ranch in sparsely vegetated land requires tens of thousands of acres, what about the native mule deer, bighorn sheep, or mountain lions that compete for feed and water? And what attention should be given to endangered species or introduced species like wild horses? The decisions became more complicated with the consideration of recreation and wilderness.

As multiple-use management was implemented, political differences were magnified. Some groups, believing government was no longer responsive to their interests, turned to various forms of civil disobedience. In 1979, with protests spreading, several western states mutinied, initiating a movement dubbed the "Sagebrush Rebellion."[9]

Politicians, including newly elected president Ronald Reagan, offered their support for the rebels. But the legal theories regarding such actions proved to be without foundation, and Congress refused to acknowledge the states' assertions. The Reagan administration split over policy. Some officials preferred to work with the states in managing lands, while administration economists promoted a real property initiative, advocating privatization. Without clear direction, the rebellion waned. But for resource users and conservative activists, the matters have never been resolved.

A key factor in the disputes is the ideological gap between the nation's major political parties. The Democratic Party has embraced environmental issues and defended government land policies. The Republican Party, proclaiming their support for growers, producers, foresters, and miners, has promoted resource expansion and sought to remove what they consider to be overly restrictive environmental regulations. The League of Conservation Voters' scorecard for congressional votes showed that in 1973, Democrats scored 56 percent on environmental reforms, while Republicans scored 27 percent. Through the years, the Democrats' score rose, while the Republicans' decreased. In 2004, while Democrats were scoring 86 percent, the Republicans' score was 10 percent. By 2015, the Democratic leadership in the Senate was averaging 99 percent, and Republican leaders averaged 2 percent. In the House, Democratic leaders averaged 92 percent, and Republican leaders averaged 0 percent.[10]

Public lands can be administered by the federal government or by state governments, or the lands can be sold to private interests. There are three readily apparent advantages to keeping lands under national jurisdiction: it allows a wide variety of uses, there is money available to manage them, and the federal government can regulate across state lines. This last factor is especially important in dealing with problems regarding water and air, which are not contained within state boundaries.

The chief concerns raised in opposition to centralized administration are the costs, the size of the bureaucracy, and specific decisions that are a detriment to particular land users. Arguments for conveying the land to the states emphasize putting decision-making closer to the people most affected. It is posited that in state management, costs could be better controlled with less bureaucracy, and there would be an increase in state income.

A threat in pursuing state ownership is that eventually, lands will be privatized. Those who fear such outcomes point out that moneyed interests, including large corporate ranchers and extractive resource developers, would wield more influence in the smaller settings than they do in the national legislative bodies. Beginning in the nineteenth century and continuing to the recent past, states have demonstrated an inability to retain lands.

At statehood in 1864 Nevada was granted 3.9 million acres in school land grants. In 1870 the state approached Congress with a proposal to trade those lands for 2 million acres of river valleys and land close to population centers. Congress approved, and the action reduced state holdings by nearly 50 percent. Subsequently, Nevada's grant lands were virtually all patented to private owners. Idaho has sold half the land it received at statehood to private interests, most before 1940, although since 2000 more than 100,000 acres have been sold. A report in 2015 showed that Arizona receives roughly half of its state trust revenue from public land sales and commercial development.[11]

Another impact of state ownership is that user fees are significantly higher on western states' public lands than federal lands. Trust lands, the most common form of western states' holdings, are required to secure the maximum financial return for schools and endowment beneficiaries. In 2013, while grazing fees on federal lands were below two dollars per animal per month, state fees ranged from just under three dollars in Arizona to a nearly prohibitive eleven dollars in Montana.

Another troublesome point is that the costs to suppress and fight wildfires, which—owing to several years of drought in the Far West—have increased exponentially, would severely impact the states. A 2016 study found that annual firefighting costs have averaged $1.8 billion annually over the past five years. States would incur significant expense in replacing the BLM and Forest Service suppression infrastructure, and one catastrophic fire could break a state budget. If management and firefighting expenses exceeded revenue, the state would face difficult choices. It could raise fees, raise taxes, or sell the land.[12]

Arguments favoring the privatization of land include that the bureaucracy would be dismantled and the money raised could be applied to reducing the national debt. Steve H. Hanke, a member of Reagan's Council of Economic Advisers, points out that "nationalized lands represent assets that are worth trillions of dollars, yet they generate negative cash flows for the government." Hanke wants the lands sold. Transferring lands to the states, he argues, is merely trading one socialist enterprise for another.

In 2015 leading Republican presidential candidate Ted Cruz discussed the possibility of land sales with Senator Mike Lee of Utah. Lee pointed out that the value of all federal land was $14 trillion, the same amount as the national debt at that time. Cruz commented, "That suggested to us an obvious and rather elegant solution for eliminating the debt and moving as much land as possible—other than national parks—into private hands."[13]

Theorists advocating privatizing lands postulate that some entrepreneurs would preserve natural wildlands to cater to those willing to pay. And it is suggested that environmental interest groups might purchase some wilderness areas to keep them from being developed. On the whole, though, privatization would drastically reduce public access.

A further concern regarding privatization is that without government management, it would be impossible to protect a region's ecology. Historically, commodity developers have been concerned with expediency in removing resources rather than methods that safeguard the environment. The evidence includes clear-cut forests, mercury from mining left in streambeds, oil-spill disasters, strip mining, mountaintop removal, and well over one thousand Superfund cleanup sites.

The points of contention in land transference bring the discussion back to federal management. Because the government has stabilized and generally balanced land use, twenty-first-century quarrels involve claims that government policies are burdensome or economically oppressive. The cyclical nature of the arguments features questions of states' rights and disagreements among the branches of government. The seeds of the conflicts were planted in Philadelphia in 1787.

1
Creating the Federal Domain

The founders who created the U.S. government spent months debating issues that remain contentious in the western land rebellion. Concepts taken from other governments and ancient institutions vied with radical openness and the promotion of freedom to define the new Republic. Hamilton, Madison, and those seeking a strong national union were confident that a constitution could allow for an effective central government. They contended that "the ideals of human rights and rules by the people required not suspicion of government but use of it." Patrick Henry, George Mason, and others were leery of an expansive government. They advocated a loose association of sovereign states where rulers and the ruled knew and trusted each other. In a strong central government, they saw threats, and their arguments led to the institution of the Bill of Rights.[1]

In 1832 the South Carolina Ordinance of Nullification challenged federal authority. Their officials argued that tariff acts were unauthorized by the Constitution and thus violated its intent. President Andrew Jackson challenged the concept that a state might choose which of the Union's laws it had to follow. South Carolina had announced that any effort to enforce the tariff would absolve it from maintaining political connections with the other states. Jackson contested the notion, saying that when the states joined together, "[they] cannot from that period, possess any right to secede, because such secession does not break a league, but destroys the unity of a nation." Congress authorized the use of force against South Carolina before a compromise was worked out. The argument foreshadowed the greater test a quarter century later, as the Civil War answered the Confederate states' attempt to secede.[2]

During the Civil War, Abraham Lincoln, the first Republican president, eliminated slavery, expanding the scope and authority of the presidency. Some fifty years later, Republican Theodore Roosevelt termed his goal on social issues "a constructive radicalism, just as it was under Lincoln." It was a philosophy he followed in formulating his land policy as well.[3]

During the nineteenth century, American forests were thought to be inexhaustible. But timber was the universal building product, and, over time, it became apparent that it was consumption that was limitless. Woodlands around the Great Lakes, in the Rocky Mountains, and in the Sierra Nevada were devastated by massive areas of clear-cutting. Unchecked, the onslaught would precipitate a dire shortage. In attempting to avoid a crisis, Congress passed the U.S. Timber Culture Act of 1873. Capitalized investors immediately found ways to contravene it.

Logging operations in the Sierra Nevada were typical. Shills, including sailors on leave, would buy the legal limit of 160 acres of forestland and immediately resell it to logging companies. William A. J. Sparks, U.S. General Land Office commissioner, later commented that government officials "appear to have been the only persons in the vicinity who were ignorant of the frauds." Lumber companies contaminated the waterways, clear-cutting to their banks and dumping sawdust and chemicals into them. Congress received numerous petitions from downstream water users entreating it to create land reserves to remedy the misuse.[4]

Attempting to salvage the timber policy, lawmakers enacted the Forest Reserve Act of 1891, authorizing the withdrawal of land from public use. Benjamin Harrison removed thirteen million acres from the market as forest reserves, Grover Cleveland removed twenty-one million acres, and in 1897 administrative control of the lands was codified.

In 1901 the newly inaugurated Roosevelt took unprecedented action. Looking to preserve the health of waterways, rangeland, and parks, as well as forests, he began withdrawing millions of acres from the market. His actions reignited the battle over states' rights. Political conservatives, who feared government overreach, together with western lawmakers, seeing their states' lands being locked away, protested. The issues they raised, including the expansion of the national government and the extent of its power, would be central to all succeeding debates over public lands.

The president's adversaries included perhaps the most dominant speaker of the House of Representatives in U.S. history, "Uncle Joe" Cannon. The diminutive and pugnacious speaker from Illinois ran the House with an iron fist. As chair of the Rules Committee as well as Speaker, he controlled the country's lawmaking. He then complained that his adversary Roosevelt had "no more use for the Constitution than a tomcat has for a marriage license." Representatives of the large sheep and cattle ranches and a number of western newspaper editors concurred, some comparing the Roosevelt administration to the Russian government.[5]

In the late nineteenth century, chaos and anarchy were the ruling forces on large parts of the West's open range. The lands were ungoverned and overstocked. "Class was arrayed against class—the cowman against the sheepman, the big owner against the little one—and might ruled more often than right."[6] In some places cattle companies, financed by Europeans or eastern banks, fenced public rangelands. In others they bought up all the land bordering rivers. With competitors unable to bring herds to water, the monopolistic companies were able to tie up millions of acres of public grazing land for free.

At the same time, giant companies were attempting to monopolize hydroelectric energy sites. But as Roosevelt withdrew areas from the market, state officials bemoaned stunted revenues. They complained that eastern politicians were retarding western prosperity. Previously, public lands had been available for common use until citizens acquired title to them. Now in the West, the federal government was eliminating unrestricted use and removing the possibility of private ownership. The policies were deemed "iniquitous" and "pernicious." A local Colorado politician declared, "We are aliens in our own country."[7]

In 1900 in *Stearns v. Minnesota*, the Supreme Court had recognized the federal government's jurisdiction over undesignated property, declaring, "[A state] shall never interfere with the primary disposal of the soil within the same, by the United States, or with any regulations Congress may find necessary for securing the title in said soil to *bona fide* purchasers thereof."

The fact that the court was identifying Congress as the proper agent of federal actions did not deter Roosevelt from claiming authority.[8]

The president's representative in his land dealings was Gifford Pinchot, chief of the Division of Forestry who became chief of the Forest Service at its creation in 1905. Pinchot was a towering, bushy-mustachioed outdoorsman who had been nicknamed "Apollo" when he attended Yale in the 1880s. He secured his friendship with Roosevelt on their first meeting when they engaged in a friendly boxing match. Pinchot declared that he "had the honor of knocking the future president off his very solid pins."[9]

The forester had studied in Germany, and he espoused a technocratic utilitarianism, managing forests for the good of all society. Early on he was supported by the naturalist John Muir. But in the late 1890s, the two split over the meaning of conservation. While Pinchot defined it as the wise use of forests as a resource, Muir believed the term should be defined as preservation.[10]

While developing management systems and training civil servants, Pinchot was outspoken in calling for tougher enforcement policies. He pointed out that the penalty for stealing a horse in the West could be death, while stealing public trees did not even cast a "shadow on the reputation of the thief." For Pinchot's efforts western newspapers variously labeled him "un-American," a "brigand," and an "absolute dictator."[11]

Anticipating the critical value of public relations, Pinchot countered the personal attacks with a nationwide campaign to influence opinions. Creating an in-house communications bureau, the forester eventually built a mailing list of 670,000 names that included newspaper and magazine editors. He stressed the importance of controlling publicity, advising district rangers that they should "use the press first, last, and all the time."[12]

A significant factor creating discontent with Pinchot's program concerned the professionals brought in to manage the lands. American associations and societies had begun establishing criteria and training for practitioners in fields such as law, the sciences, and economics. In business the managerial revolution was under way. Just as large enterprises run by salaried managers replaced small, traditional family firms, Pinchot brought in college-trained "experts." Disputes arose when civil service employees began administering lands previously the exclusive domain of stock growers and the resource industries.

Ranchers trusted their own practical knowledge rather than the rangers' education. The federal agents were thought of merely as easterners. Colorado's senator H. M. Teller said the rangers had "no conception of their duties" and were "without the ability to discharge them." Congressman Hershel Hogg was more descriptive, calling the rangers "goggle-eyed, bandy-legged dudes from the East" and "sad-eyed, absent-minded professors and bugologists."[13]

Enacting forest policy in July 1901, Pinchot shocked western land users by showing up at Cripple Creek Opera House, at the base of Colorado's Pikes Peak. Cripple Creek was a rough-and-tumble town where rejection of the new forestry laws was a matter of course. The occasion was the Trans-Mississippi Commercial Congress. The stockmen and loggers attending believed that land withdrawals and Pinchot-initiated regulations would severely impact their livelihoods. Pinchot's speech on that occasion was an abject failure, as the crowd remained unresponsive. Still, the fact that he had faced them earned him a measure of grudging respect.

A month later, Pinchot spoke in Denver, using stereopticon views to illustrate his speech about grazing on public lands. In the nineteenth century, the public domain consisted of millions of acres of "free grass." Now, he insisted, the degradation of the land required the immediate extension of land reserves throughout the West.[14]

Some ranchers conceded that government control was needed, as it was common for migrating herds to use water pumped or piped to the surface by the ranchers. Still, they did not want to be charged for using public lands. Pinchot replied that they needed to help pay expenses, since the government was allowing them to graze and protecting their rights. By 1905 a survey of Colorado cattlemen substantiated that management was needed, as 1,000 of those asked favored government control, while only 187 opposed it. Within eighteen months, as more lands were withdrawn and fees collected, the numbers were reversed.[15]

One of the major objections of westerners was that the new policies had not been authorized by Congress. The Forest Service had been created earlier that year, and grazing on public lands had been restricted by a mere forestry mandate. The administration was liberally interpreting the Constitution's Supremacy Clause, that federal laws take precedence whether conflicts result from the actions of Congress, the courts, or administrative agencies. To the current day that reading is debated by those who want the clause strictly limited.

Regardless, accusations against Pinchot did not generally utilize a constitutional argument. One newspaper editor accused Pinchot of graft and urged users to refuse to pay the fees. Another used a mocking tone: "If one wishes to read a work quite as humorous as the comic almanac he should take up the rulebook which Bub Pinchot promulgated for the government of his forestry reserves."[16]

In June 1907 the antagonistic land issue came to a head. Oregon senator Charles Fulton had attached an amendment to the agriculture bill that would stop presidential activism in the northwestern states. It prohibited the expansion or creation of reserves in Oregon, Washington, Colorado, Wyoming, or Montana.

Colorado senator Thomas M. Patterson said the amendment was like shutting the barn door after the horse had been stolen, but at least the administration would not "have the audacity to attempt to set apart any more lands." He misjudged their determination. When Congress passed the bill, Roosevelt had eight days to sign it in order to secure the funds needed for government operations. Pinchot and his staff did paperwork around the clock. By the time Roosevelt signed the agriculture bill, thirty-two new proclamations for reserves, covering sixteen million acres, had been issued for the five states. The president and Pinchot had again imposed their will.[17]

In his annual message to Congress, in December 1907, Roosevelt emphasized conservation. Calling the range "wasted by abuse," he also pointed out that the "present annual consumption of lumber is certainly three times as great as the annual growth."[18]

In May 1908 Pinchot, under the president's sponsorship, brought a large number of conservation supporters to address the Proceedings of the Conference of Governors in Washington, DC. Many of the speakers called for federal control. Dwight B. Heard, president of the Arizona Cattle Growers, spoke of stockmen wanting federal action to reduce feuds and stop rustling. H. A. Jastro, a California county official, spoke of sheep growers needing a hundred armed guards when they moved their stock on "the free public lands of the United States."

The resultant "Governors' Declaration" urged an extension of forest policies and asked Congress to enact laws to prevent monopolization of the country's resources. But, with Uncle Joe Cannon's "old guard" in the

House of Representatives following a "not one cent for scenery" policy, Congress was unresponsive.

Still, by the end of his presidency, in 1909, Roosevelt had moved land management away from private interests, corporations, and the states to the federal government. His freewheeling utilization of power, along with Pinchot's building of an expansive bureaucracy, ensured that philosophical differences regarding lands would carry forward.[19]

2

Interior Battles

A severe disagreement among Republican Party factions erupted early in William Howard Taft's administration. Taft, the "amiable giant," was Roosevelt's handpicked successor. He was said to tirelessly pursue matters until they were settled, and so it seemed he would be perfectly suited to complete Roosevelt's unfinished work. Instead, Taft's differences with his predecessor surfaced.

The most significant of the intraparty conflicts concerned the use of public lands. It burst forth in the Ballinger-Pinchot affair that involved recurring issues: executive power, the imposition of federal rules on state jurisdictions, and the availability of federal lands' resources for corporate use. But power shifted as the new administration reined in, and in cases reversed, executive actions, seeming to favor business interests.

After pledging to keep Roosevelt's cabinet intact, the new president decided Secretary of Interior James Garfield Jr. was too much influenced by Pinchot and replaced him. Taft named Richard Ballinger, formerly a reform mayor in Seattle and Roosevelt's director of the General Land Office, to the position. Neither Garfield nor Pinchot trusted that Ballinger would serve the public interest. Pinchot dismissed Ballinger as "a stocky, square-headed little man who believed in turning all public resources as freely and rapidly as possible over to private ownership."[1] Over the three months remaining in Roosevelt's term, Garfield withdrew three and a half million acres from the market.

Roosevelt had operated under the concept that on matters of the public domain, the executive could proceed unless forbidden to do so by law. An example of his unorthodox actions involved the Grand Canyon. In 1908 it was listed as a forest reserve, but a land speculator sought to control tourism by making mining claims at the area's premier overlooks. There was no time to make the canyon a national park, so Roosevelt signed a proclamation naming it and its rims a national monument. The Antiquities Act, approved two years earlier, gave the president authority to preserve objects of historic interest. But the act stipulated that monuments were to be kept to the smallest area possible. Critics protested that the entire Grand Canyon was certainly not intended to be covered under the act.[2]

Diametrically opposed to the Roosevelt philosophy, Taft and Ballinger believed action could not be taken unless expressly authorized by Congress. Within six weeks of his appointment, Ballinger restored for potential sale more than three million acres of withdrawn land. Pinchot, who had been traveling in the West, rushed to Washington, DC, to protest to Taft. After two days of Pinchot's remonstrations, the new president called in his interior secretary and ordered him to again withdraw the lands.

By early May the press was scrutinizing Ballinger's actions. A *Los Angeles Herald* headline emphasized the break with the previous administration: "Ballinger Reverses Policy of Garfield." The same week a leading San Francisco paper's editorial called attention to the fact that Ballinger had placed the three million acres on the market without advertisement, "giving a tremendous advantage to the big water power combinations." The paper concluded, "Obviously, Mr. Ballinger will bear watching."[3]

Leaked confidential negotiations over Lake Tahoe, a valued national resource, became a flash point. The Bureau of Reclamation, under Ballinger's Interior Department, was proposing to award an electric company unprecedented rights to the scenic lake's water. The *San Francisco Chronicle* decried the agreement, proclaiming in a headline: "Secret Deal with U.S. Puts Tahoe in Syndicate's Clutch." Ballinger gave his assurance that government interests would be protected, but he refused to make public the terms of the contract.[4]

In fact, the terms were egregious. In purchasing the dam on Tahoe's only outlet, the syndicate was to be given a perpetual franchise, allowing it to construct reservoirs, powerhouses, roads, and transmission lines on any public land of the watershed. The deal also allowed the company the right to bore into the lake wall fifty feet below its surface—creating a second outlet to supply an additional reservoir. Incredibly, the government would also agree to pay any damages sought against the syndicate by other river users.

California's U.S. representative William Kent, an avid conservationist and a friend of Pinchot, pressed for a congressional investigation of the proposal. California raised the issue of states' rights. One exasperated newspaper editor commented on the state's exclusion: "Here we find a power company and a government bureau calmly proceeding to divide the property without consulting the first party in interest."[5]

The protests seemed futile, as early in 1911 the pact was about to be consummated. The California Legislature suspended its rules to rush a resolution through both houses, urging President Taft not to enter into the contract. The California governor telegraphed a personal protest to Taft. Shortly thereafter, the controversy, which had raged for more than a year and a half, was settled by Taft, who directed officials to negate the deal.[6]

Ballinger later said that he was merely opposing regulatory bureaucracy. He felt big government did not allow new businesses the same opportunities as large, established companies. He argued that although regulations "stopped the big fish from swallowing his smaller fellow, it froze the status quo, preventing the little fish from becoming big."[7]

For Pinchot, Ballinger's policies offered grossly inadequate protections, and he pressed an attack against the secretary. In August 1909 Pinchot contacted President Taft, introducing Louis R. Glavis, a twenty-seven-year-old General Land Office investigator. Glavis accused Ballinger and three assistants of abetting fraudulent claims to Alaskan coal-mining lands.

Land laws in Alaska, as throughout the West, were intended to promote small farms and prevent monopoly, limiting claims to 160 acres. Clarence Cunningham, a developer who had worked with Ballinger in Seattle, had formed a syndicate with thirty-three investors, claiming lands to form a coal-mining company.[8]

Evidence of illegality included the discovery of a document from the

J. P. Morgan–Guggenheim Company, stating it would pay $250,000 for half the Cunningham group's stock. Glavis reported that it appeared that Ballinger had worked to ensure approval of the Cunningham claims.

Ballinger, reticent to speak through the press, instead presented the president with a ten-thousand-word response to the Glavis charges. Taft concluded that Glavis had only "specters of suspicions without any substantial evidence." The president allowed Ballinger to fire Glavis for disloyalty and making false charges. Afraid of further alienating the vast numbers of Roosevelt supporters, Taft sent a note to Pinchot, urging him not to make the Glavis issue his own.[9]

Pinchot rejected the plea, writing a letter to be read on the Senate floor. It called Glavis a "most vigorous defender of the people's interest" and suggested that in allowing him to be dismissed, the president had been misled. What followed would have been expected. Headlines blared: "Taft Ousts Pinchot for Insubordination." Suddenly, the capital was caught up in a "feverish excitement seldom before known in state or political circles." Pinchot's dismissal seemed to substantiate that Taft was opposing Roosevelt policies and was therefore "friendly to the corporations."[10]

There was the appearance of Ballinger's influence in the Alaska scandal, but a congressional investigation did not uncover any illegal activities on his part. It was, though, a political cause célèbre, and the hearings created daily news. Called before the committee, Ballinger seemed evasive. For a time, he tapped his foot in "a restless tattoo." At another point, he became indignant and refused to answer questions. Newspapers consistently denounced his performance.[11]

That summer the Republican divide widened. On a western swing to assist Senate campaigns, Roosevelt implied condemnation of the interior secretary. The former president touted a candidate by saying he was "opposed to that wing of the party headed by Mr. Ballinger." The headline in the *Los Angeles Herald* read, "Big Stick Swings in the Fight to Crush Ballinger." Other newspapers piled on, accusing Ballinger of being "the indefatigable friend of the special interests."[12]

The onslaught was debilitating, and in January 1911 Ballinger sent a letter of resignation to Taft, who held on to it until early March, when he accepted it. The president appointed Walter L. Fisher, a Pinchot associate and noted conservationist, as the new interior secretary. On June 26, 1911, news was released that Fisher had invalidated the Cunningham claims in Alaska. Typical of the reaction was the *Washington Times* editorial that included this headline: "Repudiation of Ballinger and Victory for Pinchot."[13]

Lost in the long, bitter controversy was the fact that Taft's approach to protecting public land succeeded. In 1910, at Taft's urging, Congress had passed an act providing that the president might temporarily withdraw lands from the market. Thereafter, Ballinger and his successor, Fisher, with the assurance of the law behind them, had withdrawn almost as much land as had Roosevelt's administration.

But the Ballinger-Pinchot affair had split the Republican Party. In the 1912 presidential campaign, conservative party power brokers and business supported Taft. Roosevelt, feeling that he had been cheated by them at the national convention, accepted the nomination of the newly formed Progressive Party. Nicknamed the Bull Moose Party, its symbol, an Alaskan bull moose, was intended, at least in part, to remind the public of Taft's backing of Ballinger's Alaskan coal debacle. In the general election the supporters of Roosevelt and Taft divided their vote, and the Socialist Eugene Debs garnered nearly a million votes. This allowed Democrat Woodrow Wilson to capture the White House, the second of only two Democrats to serve as president between 1860 and 1932.[14]

Pinchot would again engage in a massive battle with an interior secretary, but that would not take place until the 1930s, during the Franklin Roosevelt administration. In the interim, during Warren G. Harding's short presidential tenure, the Interior Department became immersed in scandal. Harding attacked the Wilson administration, accusing it of impairing the economy by employing "the withering hand of government operation." He favored a hands-off approach. But his faith in associates whom he appointed was repaid with treachery. "My God-damn friends . . . ," Harding complained. "They're the ones that keep me walking the floor nights!" His death from a heart attack, two and a half years into his term, came in the midst of several developing outrages, as his friends turned out to be friendly with corporate officials as well. One, Secretary of the Interior Albert B. Fall, had been bribed for the rights to two oil reserves, the rock outcropping above one giving the scandal its name, Teapot Dome.[15]

Calvin Coolidge, who was "opposed to any general extension of government ownership and control," succeeded Harding. And five and a half years later, Herbert Hoover, an energetic secretary of commerce for the previous eight years, followed. Hoover subscribed to the same creed as his immediate predecessors, describing himself as an "American individualist." Promoting a free-market economy, he preached a reduced role for the federal government, especially where it might be in competition with business.[16]

Renouncing Roosevelt's land policies, Hoover proposed that the surface rights to 173 million acres of federal land be given to the states. "Western states have long since passed from their swaddling clothes," he commented, "and are today more competent to manage much of their affairs than is the federal government." A forester commented that because government restrictions were unpopular, terminology was being bandied about "to hoodwink the crowd." Use the term *bureaucracy*, he posited, "and your case is won."[17]

Hoover appointed the Hoover Commission on Public Lands, headed by former interior secretary Garfield. It came up with a complex proposal to convey a portion of the federal acreage. But few or none of the states were prepared to take on the task of management, and because mineral rights were not included, the states looked on the lands as a liability. Utah governor George H. Dern commented that he "couldn't help wondering why [western states] should be deemed wise enough to administer the surface rights but not wise enough to administer the minerals." Congress viewed the plan from various perspectives, conservationists actively fought against it, and it failed.[18]

In the 1930s, when Democratic president Franklin Roosevelt reintroduced a vigorous executive-branch activism, land policy again underwent significant changes. His administration established federal management of nonforest public lands, and his New Deal infused federal money into the West. Overgrazing had contributed to the "Dust Bowl," and a committee on lands concluded that "without regulation further destruction is inevitable." Congress passed the Taylor Act in June 1934. The act established grazing districts, providing for their orderly use. By ending free access, it also effectively ended homesteading on the West's rangelands.[19]

3

Rangeland Battles

The federal bureaucracy is often denounced as an unelected, and therefore undemocratic, arm of government. Executive-branch departments and congressional agencies form the bureaucracy, administering programs the president and Congress have neither the time nor the expertise to manage. Unacknowledged in criticism of the bureaucrats is the fact that they must answer to the elected officials. At the same time, they need to be responsive to public and private interest groups. If an agency is to survive, it must obtain enough support from nonopposing "constituents" to offset those opposing it. Harold L. Ickes, the longest-serving interior secretary in history, had occasion to battle entities inside and outside the bureaucracy.[1]

In July 1933 President Roosevelt, responding to unprecedented unemployment, appointed Ickes to oversee the newly created Public Works Administration. The exceptionally capable Ickes continued with his Interior Department duties as well. The month before, he had prompted the writing of a bill to allow him to regulate grazing on public lands. Over the sixty years since the 1870s "beef bonanza," the public range had become more and more depleted. Parts of it were now typified by gullies, weeds, and dust. Ickes complained, "Some of the Western states are opposed to [a grazing bill] because their stockmen in their greed want to turn their flocks and herds onto the range without the aye, yes or no of anyone." But there were some in the West who supported the bill. Senior Colorado senator Edward T. Taylor, for whom the bill would be named, observed that the ruinous overgrazing was "eating into the very heart of the Western economy."[2]

When it appeared the grazing bill had little chance of passing, Roosevelt made it known that without it, he might have to withdraw the right of entry from all public rangelands. The threat swayed some, and, after a year of hard fighting, the Taylor Grazing Act passed. Under it each agency set aside a portion of lands for game-animal forage, with the rest dedicated to livestock grazing. Permit fees would be administered by a Division of Grazing, later renamed the Grazing Service.[3]

Officials within the administration had attacked the act as lacking sufficient controls. One aspect seemed to ensure future controversy. The act's preamble spoke of management "of the public lands pending its final disposal." It was argued that a court might interpret that phrase as giving users private property rights. And naturally, users inferred that sometime in the future, they would be able to gain private ownership. Ickes argued successfully that the bill's benefits outweighed its liabilities, including the preamble's vague implication.[4]

Farrington Carpenter, owner of a small Colorado cattle ranch and an attorney who represented stockmen, was appointed first director of the Grazing Service. Carpenter spent a year organizing sixty-five million acres into grazing districts and consulting with ranchers. He had little use for the kind of federal oversight being administered in other New Deal programs, saying, "I feel like most westerners some way or other can handle things better than these bureaus can." He concluded that, unlike Forest Service grazing—which after three decades still engendered resentment—in rangeland management local representation would help secure local compliance. He posited that "advisory boards looked like the answer." Once installed, district boards often became the determinate voices in setting boundaries, the allocation of permits, seasons of use, and the carrying capacity of the range.[5]

Carpenter's approach was antithetical to what Ickes wanted. Although the secretary did not disagree with the concept of advisory boards, he did not want to relinquish federal control. Carpenter said that Ickes "didn't have a lot of confidence in other people's integrity to make an honest judgment. . . . [Ickes] thought he should make the final decisions."[6]

While Ickes was battling Carpenter inside the department, he was also forced to stave off attacks from outside. In September 1935, acting on Ickes's advice, Roosevelt vetoed a proposed amendment to allow mandatory exchanges of state land for federal. The amendment stipulated that the land must be of equal value, but would be transferred merely on application of the state. It also proposed Division of Grazing employees be limited to in-state residents. Ickes protested that the amendment ceded too much power to the states and that the employee policy would put staff under local pressure, or "the whims of dominant local stockmen."[7]

Meanwhile, Carpenter was trying to implement the grazing law's directives. No fees had been instituted in 1935. In January 1936 a mass meeting of district advisory board members convened in Salt Lake City. They determined charges for cattle and horses should be five cents per animal unit per month (AUM), and sheep or goats should be one cent. The schedule would be implemented beginning the following grazing season.

Ranchers in Nevada immediately filed suit. They complained that the public grazing land increased the assessed value of ranch property, raising taxes, so ranchers were already paying for use. They also wanted the fact that there were substantial variations in the value of forage on different parts of the range taken into account. Nevada courts accepted the arguments. The U.S. Supreme Court did not, confirming the authority of the secretary to act as the agent for Congress.[8]

Ickes had continued to express displeasure with Carpenter's actions as grazing director. In March 1937, commenting on several cases where ranchers had appealed range managers' decisions, Ickes observed, "The cases have been kicked around [by Carpenter] until the questions to be solved have become moot." After the elections of 1938, Ickes had had enough, and he fired Carpenter.[9]

Harold Ickes was called by his wife "the most fanatical conservationist of his generation." Early in 1935 Ickes said, "I would like a chance to build up a strong conservation department and develop a strong public opinion for conservation." Meeting with the president, he requested that all conservation activities be moved to Interior and the department be renamed "the Department of Conservation." Roosevelt allowed Ickes to take the proposal to the Hill. Wallace and Forest Service leaders argued for keeping forestry in Wallace's Agriculture Department. They lobbied hard against the proposal, which passed in the Senate but stalled in the House.

To ensure Ickes would not gain further headway, Wallace brought in reinforcements in the form of Pinchot, now the governor of Pennsylvania. The battle went on for three years in speeches and over the radio airwaves. Pinchot lashed out, claiming the Interior Department's record was one of pillaging public lands and that too much centralized power boded ill for a democracy. Ickes gave as good as he got, accusing the retired forester of walking "arm and arm" with the great lumber and mining concerns and trying to keep conservation activities disunited. In the end, Pinchot's lobbying and public relations skills won out. The result kept the management agencies split, subverting the idea of conservation being managed as its own executive-branch department.[10] Ickes did not have time to agonize. War was engulfing the world's democracies, and, on the home front, he faced ongoing battles over rangeland grazing.

Nevada senator Pat McCarran's advocacy for ranchers had created recurring regulatory problems. The senator had attached two amendments to the Taylor Act in 1934 that weakened the federal government's authority: giving states jurisdiction over health and welfare issues and limiting regulators' ability to deny permit renewals.

In 1939 McCarran had redoubled his efforts. Dubbed "the political boss of Nevada," the ebullient senator was also a force to be reckoned with in Washington. Short and stocky, with a shock of white hair, in the early 1950s he would become nationally known for drafting antidemocratic laws that became McCarthyism's legislative legacy. In the 1940s he made his name on the Senate Interior Committee, opposing the bureaucracy.[11]

McCarran was the first native Nevadan to serve in the U.S. Senate, having been raised on the twenty-six-hundred-acre family sheep ranch outside of Reno. The importance of grazing in Nevada cannot be overstated. When McCarran served in the Senate, his constituency included fifteen of the fifty largest ranching operators in the country. Those Nevada spreads averaged more than nine thousand animals, almost double the average number of America's other thirty-five largest operators.[12]

When Ickes had dismissed Carpenter, the Division of Grazing was in terrible shape in terms of both its image and its finances. Getting a new start by changing the name of the agency to the Grazing Service, officials attempted to increase charges. The idea never had a chance. Fees on

Forest Service land were thirty-one cents per AUM, while the Grazing Service was charging a nickel. When the grazing bill had been under evaluation, Ickes had stressed that fees would pay only for range administration, "nothing more." McCarran now insisted that even the five cents was unjustified, since there was no evidence that the range had been improved by the agency.

McCarran argued that the western stockmen's longtime use of the land conveyed de facto ownership. But that contention seemed intended only to rally his constituents. His real efforts were directed at manipulating the land-use system, ensuring the balance of power favored the stock growers. In May 1941, to press his case, he persuaded the Senate to adopt a resolution providing for investigation of the Grazing Service. The hearings, held throughout the West, lasted into 1947, serving as a long-term public relations blitz.

At mass meetings McCarran recited figures demonstrating the immense growth of the Grazing Service, with a budget rising from $250,000 in 1936 to $750,000 in 1941. The statistics set the stage for his condemnation of any proposed fee change, as he would argue more money would mean only more bureaucracy.[13]

One of McCarran's major successes had been in promoting state advisory boards and establishing a National Advisory Board Council to augment the district boards. The trilevel concept produced a complex review system that obstructed agency proposals, further empowering stock growers.

Ickes argued that the boards did not protect small ranchers or conserve the land. They were diluting the power of Grazing Service officials, who were sworn to use "impartial and independent judgment." He also objected that the boards were intended to be "advisory" and should not make executive decisions. But his protestations had little effect in the West where determinations were being made. Critics claimed the national council's overpowering influence constituted a private government. Over the next six years, it blocked any proposed rate increase.[14]

In the spring of 1945 the president died, and on February 15, 1946, Ickes resigned, having served thirteen years. The loss of his adversary did not slow McCarran. He attacked a new proposal for increased funding, declaring it was put forward so that "a very few grasping individuals could satisfy their lust for power by building a bigger empire."[15]

The House of Representatives' Subcommittee on Interior Appropriations originally backed the Interior Department in calling for fee increases. Senior Oklahoma representative Jed Johnson said that everyone knew the problem: the agency had been virtually turned over to "the big cowmen and sheepmen of the West." But with McCarran, the powerful National Woolgrowers Association, and the cattlemen associations asserting pressure, the House reversed itself. Because the grazing agency was intended to be self-supporting and it was collecting so little in fees, the appropriations committee cut funding.[16]

Shortly after the budget cuts, in July 1946 the Grazing Service and the General Land Office were combined to create the Bureau of Land Management. If the action had been intended to better serve the public, it failed. Congress did not provide a clear mandate regarding the new agency's mission. Its new budget forced the dismissal of two-thirds of the agency's employees, reducing their numbers from 250 to 86. Those who remained were charged with the impossible task of managing grazing land totaling 150 million acres while sorting through thirty-five hundred statutes enacted over the previous 150 years. Realizing the potential for chaos if the lands again became completely unregulated, the advisory boards used funds appropriated for rangeland improvements to pay BLM salaries. Consequently, in the bureau's inaugural years, BLM officials were financially beholden to those they were meant to supervise.[17]

There were seventy thousand substantial ranches in the West. Only twenty thousand used the public range, but they had an inordinate amount of influence. Their leaders wanted the public land made available for purchase. The major livestock associations and two Wyoming politicians, Senator Edward Robertson and Representative Frank Barrett, chairs of their respective land committees in the U.S. Senate and House, supported the idea.

Robertson, who had previously served as vice president of the Wyoming Stock Growers Association, introduced a bill that would convey public lands to the states wishing to take over management. The president of the National Wool Growers Association argued that the bill did not go far enough. He thought the states should be required to take all federal lands. Other similar proposals were advanced. In the Seventy-Ninth Congress, between January 3, 1945, and January 3, 1947, fifty-eight bills to convey public lands were introduced and defeated.[18]

Still, low fees and minimal regulatory controls demonstrated the success of McCarran in outmaneuvering the Interior Department. His hearings, in eighteen communities, had involved listening almost exclusively to grievances from livestock owners. But a significant result of the committee's approach was that conservationists pushed back. Their tactics, as opposed to McCarran's regional approach, involved rallying a national constituency.

Western writer Arthur Carhart, who many years earlier had helped conceive the American concept of wilderness, and historian and columnist Bernard DeVoto stepped forward as the leading voices for the unfenced West. In a January 1947 *Sports Afield* article, titled "This Land Is Your Land," Carhart charged Senator Robertson with a conflict of interest because of his numerous sheep-grazing permits. Western stock raisers fought back, branding Carhart a liar and a communist, although he was a lifelong Republican aligned with the Hoover bloc. The writer continued his crusade, turning the spotlight on the intimidation of government employees. He railed against physical attacks and threats to kill officials using "the 30-30 method," that is, shoot them with a 30-30 rifle. He continued writing articles even though in answer he received an anonymous "30-30 method" threat of his own.[19]

DeVoto was also attacking the ranching elite in a monthly column in the popular *Harper's Magazine*. He called efforts to remove lands from federal ownership "one of the biggest land grabs in American history." He maintained that his readers were the rightful owners of the public domain. "The Cattle Kingdom never did own more than a minute fraction of one percent of the range it grazed," he wrote. Because grazing fees were substantially less than fees on private lands, he argued, the public was subsidizing the cattlemen. DeVoto went on to attack those promoting lands being turned over to the states because he believed the states would merely hold them until they could be disposed of in private sales. He directed specific vitriol toward McCarran, whom he accused of "stooging" for "the big stock interests."[20]

In early 1947 there was a joint meeting of the cattle and wool associations in Denver. Their leadership came up with a proposal that they buy 145 million acres of government land at anywhere from 9 cents to $2.80 an acre, with thirty-year mortgages at 1.5 percent interest. To promote their scheme, they used the misleading slogan "Return the public lands to the

West." The lands had always been under federal jurisdiction. Most western states' enabling acts, detailing terms upon which they would be admitted to the Union, included the condition that they disclaimed forever rights and title to undistributed federal properties.

In pursuit of their cause, the associations also utilized growing reactionary fear. They complained that they were at the mercy of "Communist-minded bureaucrats." J. Elmer Brock, vice president of the American National Livestock Association, asserted that most federal officials "are tinged with pink or even deeper hue." But the groups made a mistake in their privatization effort. They included national forestlands and even parklands in their proposal. Pushback came from civic groups, outdoorsmen, hunters, and conservationists.[21]

Congressman Barrett, who later served as Wyoming's governor and senator as well as general counsel in Eisenhower's Department of Agriculture, now took action on behalf of the associations. The public lands subcommittee that he chaired engaged in a McCarran-like tour of rangeland states.

At some of the meetings Barrett used "intemperate language" that "reached screaming intensity." He brought crowds of stockmen to their feet, applauding and cheering. But at several meetings city officials and ranchers with water supplies endangered by rangeland abuse spoke out. Although they were limited to brief statements, generally at the end of meetings, their concerns became part of the public record.

The national media again became involved. *Collier's* and the *Atlantic Monthly* denounced the House committee's "rigged hearings" and the "bias displayed by the chairman." The *Denver Post* spoke of "Stockman Barrett's Wild West Show." Barrett's excesses were apparent to those reading the committee's final report, including Secretary of Agriculture Clinton P. Anderson, who denied their major recommendations.

In the early 1950s the advocates of allowing private interests to buy rangeland again pushed their agenda. Backing for the stockmen came from the U.S. Chamber of Commerce, timber interests, mining and chemical corporations, big oil, and power conglomerates. The affiliation also included the Republican Party, whose 1952 platform sought to provide judicial reviews of "administrative invasions" of land users' rights. But America was changing, and legislation to extend the ranchers' dominant use was repeatedly defeated.[22]

4

Multiple Use

In World War II the United States had become the arsenal for Democratic nations. The military used the West for bases and, in the desert, bombing ranges and nuclear testing.[1] In the first half of the twentieth century, the federal government had built six hundred dams in western states, allowing the use of ten million acres of farmland. But an unintended consequence was that the dams provided for fifty-eight power plants. The power, in turn, supported spectacular growth of urban areas, making the West the fastest-growing region in the country. Additionally, utilizing leisure time, the burgeoning middle class began traveling to public recreation lands. Congress attempted to balance the competing demands created by growth. Each of their actions further squeezed the western stockmen and extractive resource interests.

John Muir had called the Sierra Nevada the "Range of Light" because one morning from across California's Central Valley, the mountains "seemed not clothed in light, but wholly composed of it." In 1892, looking to protect the range, Muir and 182 charter members founded the Sierra Club. Sixty years later, David Brower, a world-class mountain climber, became the Sierra Club's first executive director.[2]

Combative as well as zealous, in the 1950s Brower led an unprecedented battle against the BLM over major dam projects. The disputes ended with the removal of two of ten dam proposals from the Colorado River Storage Project. Significantly, the Bureau of Reclamation's final project authorization prohibited the construction of dams within national parks or monuments.[3]

During his seventeen-year tenure, director Brower sought to amplify the Sierra Club's political power. Calling to mind Pinchot's campaigns, Brower used the media, including ads and articles in magazines and newspapers, to promote causes. When areas were threatened with development or destruction, the organization encouraged recreational use in them to illustrate their biocentric value. Beginning in 1959, the Sierra Club published a series of exhibit-format books, using the work of Ansel Adams and other acclaimed photographers. The works focused on threatened areas of unspoiled nature and brought in more than ten million dollars for the club.

The group's efforts to inform public opinion and influence Congress were so effective that by 1966, the Internal Revenue Service (IRS) rescinded its tax-free status. Brower and his team used that setback to their advantage, advertising the loss in major newspapers and doubling their membership in a year from thirty-nine thousand to seventy-eight thousand. Brower commented, "People who didn't know whether or not they loved the Grand Canyon, knew whether or not they loved the I.R.S."[4]

In 1960 the Republican Party was ambivalent regarding federal resource agencies. The Dwight D. Eisenhower arm of the party, in control through his two terms in office, was under assault from a more conservative Barry Goldwater faction. Eisenhower had gained the support of a majority of the electorate by promoting a "Modern Republicanism." He declared, "I am conservative when it comes to economic problems but liberal when it comes to human problems."[5]

Arizona's Republican senator Barry Goldwater, whom historian Michael McGerr called "the John the Baptist of the New Right," challenged Eisenhower, igniting a revolution. Goldwater saw himself, above all, as an Arizonan—a product of the land. His campaign advertisements declared, "He has unrolled his bed on Arizona's wind-chilled mesas. Roamed her forests and valleys. And explored the yawning depths of her sheer canyons." Politics was, he said, "the art of achieving the maximum amount of freedom for individuals that is consistent with the maintenance of social order."[6]

Harry Truman's powerful secretary of state Dean Acheson had declared that the New Deal "conceived of the federal government as the whole people organized to do what had to be done." Eisenhower's undersecretary of labor Arthur Larson added a phrase to answer Acheson: "If a job has to be done to meet the needs of the people, and no one else can do it, then it is the proper function of the federal government." Goldwater argued that the New Deal perspective and the Eisenhower Republicanism were equally wrongheaded. He warned, "Throughout history government has proved to be the chief instrument for thwarting man's liberty."[7] Not surprising in the political realm, while persistently denouncing federal subsidies, because Arizona's growth relied heavily on federal water projects and defense spending, Goldwater fought to acquire them.[8]

In June 1960 Eisenhower demonstrated the difference between his Republicanism and that of Goldwater by approving the Multiple Use and Sustained Yield (MUSY) Act. The act directed that management of the national forests must provide equally for various resources: outdoor recreation, watershed, and wildlife as well as range and timber. A critical stipulation in the act acknowledged that multiple-use practices would not always yield the highest financial remuneration. Cheered by recreationists and conservationists, the provision was threatening to resource extractors.[9]

Upon taking office in 1961, John F. Kennedy further shifted land-policy balance, appointing Stewart Udall as secretary of the interior. Under Udall the administration created a program parallel to MUSY for BLM lands. The shift was dramatic. The Taylor Grazing Act of 1934 had specifically identified grazing as the principal use of the federal range, and grazers had dominated the BLM since its creation. The MUSY mandate was inherently corrosive to the grazers' interests.[10]

In fact, all those utilizing or extracting resources on public lands had reason to worry. In the West the population was shifting: those living in metropolitan areas rose from 43 percent in 1940 to 64 percent in 1960. In the farthest west states, Alaska, Washington, Oregon, Idaho, California, Nevada, Arizona, and Hawaii, the population in cities in 1960 was 76 percent.[11]

In 1958 Congress established the Outdoor Recreation Resources Review Commission to formulate recommendations for the country's growing recreational needs. By 1962 the ORRRC had determined that "90 percent of

Americans engage in some form of outdoor recreation." It urged a significantly higher priority for it, disparaging the BLM, which had no money allocated for recreational use. Soon the agency was giving recreation similar attention to that of range, timber, and mining. The BLM emblem, which had featured a miner, a rancher, an engineer, a logger, and a surveyor, was changed to depict simply a mountain, a tree, and a river valley.[12]

The new administration upped the ante by promoting a redefinition of conservation. President Kennedy said, "We must do in our own day what Theodore Roosevelt did sixty years ago and Franklin Roosevelt thirty years ago: we must expand the concept of conservation to meet the imperious problems of the new age."[13]

Secretary of the Interior Udall was a westerner, an outdoorsman, and an athlete, having played basketball on a championship University of Arizona team. He took an active part in the department's decision-making process. Considering the proposed Marble Canyon dam, which if constructed would flood part of the Grand Canyon, Udall rafted down the canyon and then rejected the project. He stressed the need to judge resource development with an eye to the future. He denounced "modern land raiders" for seeking short-term rather than long-term gains.[14]

In one of its first actions, the administration reorganized the BLM's state and national grazing advisory boards as "multiple-use advisory boards." In March 1961 Secretary Udall attended the national board council's twenty-first annual meeting. Declaring that, along with grazing, other uses of the public range now needed to be considered, he appointed himself cochair of the board. He also changed its makeup, expanding wildlife and adding resource users and other interest group representatives. The board, previously composed of twenty-three members (ten cattle representatives, ten from the sheep growers, and three from wildlife), by 1967 was composed of forty-two members (ten each from the cattle and sheep industries, ten others from wildlife, and twelve at-large members).[15]

Threats to commodity users multiplied, as Congress deliberated creating lands as wilderness areas, "where the earth and its community of life are untrammeled by man." The designations would obviously reduce resource-usable public acreage, and there was strident opposition from western industries.

The Sierra Club had begun sponsoring biennial wilderness conferences in 1949. In December 1960 the future Pulitzer Prize–winning author

Wallace Stegner composed a letter to the ORRRC in support of wilderness designations. Stegner's message was that wilderness is "a spiritual resource." Describing Robbers Roost country in Wayne County, Utah, he wrote: "It is a lovely and terrible wilderness, such a wilderness as Christ and the prophets went out into; harshly and beautifully colored, broken and worn until its bones are exposed, its great sky without a smudge or tint from Technocracy, and in hidden corners and pockets under its cliffs the sudden poetry of springs. Save a piece of country like that intact, and it does not matter in the slightest that only a few people every year will go into it. That is precisely its value." For those who could not get there, he proposed that "they can simply contemplate the *idea,* take pleasure in the fact that such a timeless and uncontrolled part of earth is still there."[16]

Secretary Udall used the letter as the basis of a speech to a wilderness conference in San Francisco, and the Sierra Club printed it in their report on the session. The *Washington Post* published it, and the writer himself included it in a book of his essays. It soon traveled around the world, being illustrated and produced in posters in such distant locales as South Africa and Australia. The Sierra Club used a phrase from the letter, "the geography of hope," as the title for an Eliot Porter book of photographs. The letter had its intended effect, as it bolstered ORRRC findings, and the commission recommended enactment of legislation that would establish wilderness areas.[17]

The Wilderness Society, founded in 1935, had been the prime advocate lobbying to create a wilderness act. The first federal wildland preserve, New Mexico's Gila Wilderness, was set aside because of the efforts of forester Aldo Leopold in 1924. Leopold later became one of the five founding members of the Wilderness Society and devoted much of his life to preserving the West's untamed lands. His pioneering concept of a "land ethic" stressed that humans are part of a community that includes soils, waters, plants, and animals. "In short," he concluded, "a land ethic changes the role of *Homo sapiens* from conqueror of the land community to plain member and citizen of it."[18]

In 1956 Howard Zahniser, the Wilderness Society's executive director, initiated the eight-year congressional battle that led to the review of sixty million acres for possible protection. Zahniser's attributes included a friendly demeanor and idealism tempered by the ability to compromise. Rather than argue against economic progress, he appealed to the patriotic concept of a healthy country. Preserving wilderness is not fighting progress, he contended; it is generating a force that might be renewed generation after generation, "a wilderness forever." He secured support from national women's groups, civic organizations, and labor. He was a moderating voice, accepting the continuation of grazing in designated areas of wilderness in perpetuity, mineral exploration until 1984, and commercial guided trips and camps.

Throughout the process, Zahniser, battling ill health, never missed a hearing for the bill. There were sixty-six rewrites of the act before it passed in the Senate 73–12 and in the House 373–1. Generally working on a white tablet at his kitchen table, Zahniser had produced the final version, although he died four months before it became law. In September 1964 the act, protecting some nine million acres, was signed by President Lyndon Johnson.[19]

The Wilderness Act was championed, in large part, by Democrats, although Pennsylvania Republican John Saylor was a leading proponent in the House of Representatives. The maverick Saylor was serving the interests of the private coal mines in his rural district by removing other land as possible competition. But he was committed to protecting the environment as well, later helping secure passage of the National Scenic Trails Act.[20]

Another maverick was the chief opponent of the Wilderness Act in the House of Representatives, Colorado Democrat Wayne Aspinall. Like Senator Goldwater, he supported government's economic involvement in reclamation projects in the West. But as the chair of the House Interior and Insular Affairs Committee, he was committed to guarding states' rights in developing their resources. Despite the wilderness bill gaining seventeen hearings and twice being approved in the Senate, Aspinall kept it locked up in committee year after year.[21]

Pressure from President Kennedy finally brought the debate out in the open. At the White House Conference on Conservation in 1962, Aspinall, who for eleven House terms represented a district of ranchers, farmers, miners, and loggers, challenged Kennedy's redefinition of conservation. He stressed that although the "purist preservationist group" had usurped the term, the original definition had to do with preventing waste while using resources.[22]

Aspinall's ally in the Senate was Goldwater. The Arizonan supported some environmental issues—for example, backing legislation to increase the size of the Grand Canyon National Park. But his faith in capitalism and mistrust of government generally predominated, and he voted against the Wilderness Act, arguing that it would lock up resources and deny tax revenue to the states.[23]

In the 1960s women's activism and student and countercultural protests were instrumental in creating an environmental ethic. Rachel Carson wrote *Silent Spring*, a 1962 best seller; the League of Women Voters pursued a clean-water campaign over a number of years; and three wives of the University of California–Berkeley faculty organized the Save the San Francisco Bay Association. During the second half of the decade, college students across the nation held environmental teach-ins and formed eco action groups. By the 1970s the environment overtook Vietnam as the number-one campus issue.[24]

Passage of the Wilderness Act was a turning point in the sixty-year land-management policy battle. Completely removing sites of potential oil, timber, and mineral extraction from the areas produced charges of hidden intent. Resource users asked how much more would be removed. Distrust turned to indignation, as through the years wilderness acreage expanded from nine million to more than one hundred million acres.[25]

Stewart M. Brandborg, a former Idaho Fish and Game official, followed Howard Zahniser as executive director of the Wilderness Society. After passage of the act, any proposed wilderness area had to gain approval through a separate congressional review. There were innumerable potential wilderness areas. The society needed to become informed in order to promote their designation. Brandborg described the effort: "I started traveling with a stack of membership cards, from Montana to Arizona, all the old Frontier Airlines states. I'd call ahead and say I want to talk about the wilderness law—can we meet for breakfast or lunch, in your home? It needed ongoing leadership."

Brandborg was successful in stimulating public participation in hearings. In the end, 134 additional wilderness areas were added, more than the agencies requested. The protection of the undeveloped expanses gave resonance to the West's romantic image as America's last frontier.[26]

When Lyndon Johnson succeeded to the presidency, things became even more complicated for resource interests. The new president was not merely following the Kennedy conservation policies; one of Johnson's stated goals was to surpass the conservation achievements of Franklin

Roosevelt. He retained Udall in his cabinet. To commodity developers' dismay, Udall declared, "Our mastery over our environment is now so great that the conservation of a region, a metropolitan area, or a valley is more important, in most cases, than the conservation of any single resource."[27]

In February 1965 Johnson delivered a "Special Message to the Congress on Conservation and Restoration of Natural Beauty." He asserted that "nature is nearly always beautiful." And he declared that, while beauty is difficult to define, "we do, for the most part, know what is ugly." Commodity users took this statement to imply that disruptive activities, like timbering, drilling, dam building, or grazing, were ugly.

Between 1966 and 1968, Representative Aspinall held numerous public meetings intended to influence Congress. He was hoping to ease antidevelopment regulations or simply assign ownership of public lands to the states or existing users. His advocacy had little effect, as Congress, in large part, followed the Johnson agenda.[28]

Along with water- and air-quality acts, Johnson's environmental legacy includes signing the National Trails System Act and the Wild and Scenic Rivers Act of 1968. By the end of Johnson's administration, the wilderness system was expanded by 800,000 acres and 2.4 million acres were added as national parkland. In total, during his presidency he signed almost three hundred conservation and beautification measures. The environmental movement was on a decadelong winning streak. For commodity interests, there was no room for optimism until Johnson announced he would not run for reelection.

Richard Nixon had left environmental issues to the Democrats in his earlier presidential run. His election in 1968 and his appointment of a self-described development advocate, Alaska governor Walter Hickel, as interior secretary buoyed the spirits of the prodevelopment faction. Major newspapers, liberal senators, and environmentalists raised an outcry. They painted Hickel as antienvironment. A change in public land policy seemed at hand.[29]

5

The Radical New Conservation

By propagating the notion that we are part of the land community rather than masters of it, Aldo Leopold set the stage for the environmental movement. In the battle over regulating land use, environmentalists looked to the land ethic rather than economics. Commodity extractors were likely foes. Although long-term users like ranchers generally supported careful land management, other extractive interests prospered by emphasizing expediency. Profiteering resulted in stripped hillsides, contaminated soils, and polluted waters.

A growing awareness of environmental damage, along with calls for protection of settings for recreation, stimulated a change in the definition of the term *conservation*. By the 1960s professionals in federal land agencies and environmentalists had reappraised the Gifford Pinchot definition. Conservation for Pinchot had meant resources would be developed in a way that ensured sustainability for future generations. It was now generally supplanted by John Muir's concept of conservation as preserving nature.[1]

The era's social, cultural, and political upheaval was splintering America along ideological lines. The Cold War intermixed with momentous events, including the Vietnam War; the assassinations of Medgar Evers, the Kennedys, and Martin Luther King; demands for civil rights; the counterculture of young people "dropping out"; and mass demonstrations, some of which turned into riots. Early in the sixties Americans believed the federal government could solve the country's big problems. That notion included multiple-use land policy replacing dominant use. But the push for far-reaching changes, in conjunction with war's discord, created political backlash in conservative circles beginning in the West.

Goldwater had set the tone for New Right westerners in the early 1960s by attacking "elite easterners" and big government while touting individual initiative and the primacy of states' rights. Goldwater embraced all things western. Tenaciously forthright, he projected a tough-guy demeanor, speaking at times like the leading man in a Hollywood western. Unfortunately for his presidential ambitions, with his comment during the Cold War that "extremism in the defense of liberty is no vice," he could also sound like a dangerous gunslinger. After losing the presidential race in 1964, he infused his blunt choice of words with humor, saying of himself, "If I had to go by the media reports alone, I'd have voted against the sonofabitch, too."[2]

A Goldwater disciple, Ronald Reagan, in his run for the governorship of California in 1967, became a leader of a populist movement. At the time, America's political culture allowed claims of affiliation with the heroic mythical West to substitute for personal history. Reagan had moved from Illinois to Hollywood to become a star in western B films. He was not a cowboy, but he often wore a cowboy hat, conflating his western image with his role as governor and later as president. Those around him burnished the popular conception of him as the good guy: Goldwater without the hard edge and verbal gaffes.[3] In his campaign for governor, Reagan listed outrages, including disrespect for authority, a morality and decency gap, unjustified taxes, and the Johnson administration's handling of Vietnam, where he proclaimed the troops "are being denied the right to try for victory."

Richard Nixon continued the Republican populist campaign. After becoming president in 1969, he deflected criticism of his escalation of the war by attacking protesters as "bums." His soon-to-be-disgraced vice president, Spiro Agnew, labeled the young people "un-American."[4]

Surprising to many, an area of noticeable achievement during the Nixon presidency was the establishment of protections that form the foundation of today's environmental programs. The quantity and scope of his administration's actions are unparalleled in the modern era, as he managed to bridge the gap between protections and resource use.[5]

After the Santa Barbara oil spill, in which a ruptured well poured oil into the ocean for ten days, Nixon publicly declared 1970 to be the year of the environment. In April twenty million Americans rallied across the country for the first Earth Day. During the next two years Nixon pioneered laws that enhanced environmental protections in land-use policies. The progress was made despite the increased promotion of extractive resource interests that had begun hiring former members of Congress as lobbyists.[6]

In one of his most radical acts, Nixon moved 5,650 government employees from other agencies, and $1.4 billion from other programs, to create the Environmental Protection Agency (EPA). Working with Congress, he signed the Resource Recovery Act, governing the disposal of hazardous waste, and proposed legislation that regulated mine sites, inhibited commodity extraction in coastal wetlands, and strengthened protection of endangered species. And he approved legislation expanding the Clean Air Act's federal mandate.[7]

The Clean Air Act of 1967 had generally left it to the states to set standards that would be approved by the federal government. But various approaches had created inconsistent standards and delays. In a move that would garner little credit from other Republicans, Nixon moved away from states' rights, calling for the federal government to set the air-quality standards and oversee state implementation.[8]

He made a similar decision regarding off-road vehicle use. The vehicles had been identified as a threat to public lands, and the Interior Department concluded their use should be regulated by the states. Nixon instead directed federal land agencies to establish criteria to determine if areas and existing roads should be open or closed. Within five years the Carter administration amended the order to tightly regulate ORV use anywhere they were having adverse effects.[9]

On January 1, 1970, Nixon had signed into law the National Environmental Policy Act (NEPA). It mandated that any project using federal funding must prepare an impact statement describing its effects on the environment. The act opened government actions to the scrutiny of environmental groups that could challenge public land use in the courts.

Challenge they did, beginning with the Trans-Alaska Pipeline, which

Interior Secretary Hickel began attempting to push through just as NEPA went into effect. Land disputes between Native peoples, the state of Alaska, and the federal government had caused Secretary Udall to halt consideration of the eight-hundred-mile hot-oil pipeline in 1968. When the route was classified in the *Federal Register* as a utility and transportation corridor, Hickel lifted the freeze.

Rogers Morton, chair of the Republican National Committee who would soon replace Hickel as interior secretary, urged that the project move forward as a means of maintaining Republican control of the Alaskan political estate. The secretaries of defense and state also urged construction, saying the pipeline was necessary for national defense.

Hickel issued an 8-page environmental impact statement regarding the transportation corridor. The statement concluded that the road would pose no threat to the environment. Three environmental groups immediately filed suit, questioning the thoroughness of the statement and saying the road and the pipeline needed to be evaluated together. The district court judge found for the plaintiffs, enjoining the project until a more comprehensive statement was produced.[10]

The second draft, 246 pages, completed in January 1971, did not make it to court. Objections from the environmentalists were seconded by at least one BLM official and former secretary Udall, who said he was "deeply disturbed by its glaring omissions." The Interior Department's solicitor also found the document to be "wholly inadequate," concluding it would not hold up under judicial review.

Officials immersed themselves in a third version. The writing of the project eventually totaled 175 man-years of work and nine million dollars. The product was a six-volume tome, with three additional volumes of supporting material. The district court removed its injunction, but once again environmental groups appealed. Whether the decision would stand was problematic. So, with a nationwide oil crisis growing, Congress acted, exempting the project from any further NEPA requirements. The lesson learned was that agencies were now required to prove to the courts that all environmental issues had been considered before projects would be approved.[11]

In his 1970 address on the environment, Nixon advocated for an increased number of recreation areas. Previously, federal property use was determined by who got there first. Now he wanted all agencies to evaluate their lands and determine if designations should be changed or if lands should be sold to finance other more usable recreation areas.[12]

Nixon's actions on the environment reflected the mood of the country. When two Standard Oil tankers collided in San Francisco Bay in 1971, the resultant spill led to the formation of new environmental organizations. In 1972 the League of Conservation Voters formed. That same year, at the urging of members from across the country, the League of Women Voters changed its statement of principles. Formerly, the league had promoted "conservation and development"; now it removed the term *development*.[13]

Richard Nixon was an enigma. Despite all his achievements, his personal regard for the environment was tenuous. Russell Train, who chaired Nixon's Task Force on the Environment and served as the administrator of the EPA, noted that Nixon had "little personal interest or enthusiasm for the subject, [although] he and his immediate advisors realized the political significance of the environmental issue." To associates Nixon presented a more candid perspective than his public pronouncements, insisting that he would not let "nature lovers" get in the way of a strong economy.

When in conflict Nixon sided with business or industry interests, as when the Water Pollution Control Act had to be passed over his veto. And his private declarations could be startling. Speaking to auto-industry executives, he said activists were not interested in safety or clean air; they wanted to destroy the system, to "go back and live like a bunch of damned animals."[14]

At times the Nixon administration pursued the politically expedient path on land issues. On January 7, 1972, a memo was sent in secret to the Forest Service and BLM proposing to reduce the clear-cutting of forests. The secret did not last. On January 8 presidents of several forest-product companies met with Nixon's secretary of agriculture, Earl Butz, and the heads of the two agencies. Within the week an official White House statement announced there would be no executive order pertaining to reducing clear-cutting. Oklahoma senator Fred Harris commented that the action proved the old adage: "Public interests win in public; and private interests win in private."[15]

With the Wilderness Society rallying a large constituency, public pressure began building to expand wilderness areas as part of the West's legacy. Nixon spoke expansively of increasing protections, but the areas he approved came from already protected National Park Service or Fish and

Wildlife Service lands. He omitted Forest Service lands and their primitive old-growth forests that the timber industry coveted.[16]

His problems with environmentalists were generally subsumed by other, more demanding, issues. Running for the presidency, Nixon had promised to end the Vietnam War and restore law and order. Since he won the election by a razor-thin margin, the war had escalated, inflation and joblessness were rising, and civil unrest had worsened. In 1969 antiwar demonstrations shut down parts of the nation's capital and other major cities. On November 15 250,000 people marched in Washington, DC. The following year demonstrations by college students across the country included the May 1970 incident at Kent State, when panicky National Guard troops opened fire on students, shooting to death two young men and two young women. Ten days later, at Jackson State College—a historically black school in Jackson, Mississippi—a confrontation led state and city police to fire 150 rounds in some thirty seconds. Two students were killed, and twelve others were wounded.

Secretary of the Interior Hickel wrote a letter about the killings to the president that attracted worldwide attention. The always outspoken Alaskan wrote, "[History] shows that youth in protest must be heard." Nixon pursued a different course, some months later replacing Hickel with Rogers Morton, the chair of the Republican Party's political committee.

That same year backlash against protesters had led construction workers in several cities to use lunch hours for patriotic demonstrations supporting the Vietnam War. "Thank God for the hard hats," the president exclaimed. Nixon had already appealed cynically to racists, earning votes with his "southern strategy." Now his team began what would be a successful effort to lure working-class white males away from a traditional alliance with the Democratic Party. The press labeled the plan "Nixon's blue-collar strategy."[17]

Throughout the twentieth century, the Republican constituency included business interests, while Democrats took up the economic concerns of workers. Rather than a higher living standard, Nixon offered the blue-collar workers cultural recognition. The president assured them that he was on their side, pointing out that true Americans were deeply offended by hippies and the left-wing protesters who were burning draft cards, bras, and American flags. As violence in the streets increased, Nixon's team painted the Democrats supporting free speech as opposing

American values. Nixon claimed that his policies would counter the permissiveness and "lower moral standards" that led to the disruptions. A *New York Times* cartoon presented a visual representation of Nixon's effort: a pipe fitter joining a Wall Street broker in using an American flag to pummel a hippie.[18]

Although the Nixon administration's efforts institutionalized the consideration of environmental values throughout government, after resigning he inveighed against what he considered the extreme elements of the movement. He said that he had sought to find a balance between protecting the environment and economic growth. But, he added, since he had instituted the EPA and the Endangered Species Act, environmental programs had "run amuck." The Republican Party agreed, soon including environmentalists among their opponents.

6

The Sagebrush Rebellion Begins

The subject of states' rights is central to discussions about public lands. The Tenth Amendment says, "The powers not delegated to the United States by the Constitution, nor prohibited by it to the States, are reserved to the States respectively, or to the people." Some states' rightists carry small copies of the Constitution in their shirt pockets to symbolize that the Tenth Amendment proves states' rights are preeminent on land issues.

There are two places in the body of the Constitution where the states rightists' argument is countered. The Supremacy Clause, Article VI, Clause 2, is the cornerstone of the entire Constitution. It mandates that in conflicts between state and federal law, federal authority takes precedence. As regards land, Article IV, Section 3, states, "The Congress shall have Power to dispose of and make all needful Rules and Regulations respecting the Territory or other Property belonging to the United States; and nothing in this Constitution shall be so construed as to Prejudice any Claims of the United States, or of any particular State."

The leading modern court decision regarding federal property, *Kleppe v. New Mexico,* in 1976, dealt with the Interior Department's right to protect wild horses and burros on public property. The State of New Mexico claimed the federal government could act only if there were interstate issues or if public land was being damaged. In a unanimous ruling, the Supreme Court found against the state, confirming the federal government's proprietary and legislative power over the public domain.[1]

It is often proposed that federal management favors environmental concerns, while local and state governance favors resource use. This was certainly not true after environmental laws were established in the 1960s and 1970s. Some government administrators continued to advance the cause of resource extractors as if complying with the old dominant-use policies.

The Wilderness Act specifically prohibited permanent roads within wilderness areas. In 1975 *Field & Stream* featured an article, "The Forest Service versus the Wilderness Act." The writer reported that the agency was rushing to build roads contiguous to the River of No Return Wilderness in Idaho so they could sell the timber in those forests. The article said the action was driven by Agriculture Secretary Earl Butz, who was also encouraging farmers to "get big or get out." The author believed that the majority of Forest Service officials would acquiesce to Butz's directives to expand timber removal because they were "product oriented." In the spring of 1970, an internal Forest Service task force that studied clearcutting on the Bitterroot National Forest in Montana had come to a similar conclusion. It found that the agency was still working from the 1950s intensive harvesting model so that resource production came before any other consideration.

As regarded the Wilderness Act, when considering areas for designation, the Forest Service followed a purity policy: unless areas were pristine, they would not qualify. The agency was constructing the roads adjoining the Idaho wilderness, *Field & Stream* asserted, so that when the time came, officials could say, "Look, you can't include this area [as wilderness]; there are roads everywhere."[2]

It was a recurring story. In 1965 the Forest Service had constructed a road into East Meadow Creek, a primitive area eight miles north of Vail, Colorado. They intended to allow a large timbering operation, planned in the late 1950s, to go forward. In 1969 preservationists filed suit, contending East Meadow Creek was designated as primitive and was contiguous with higher elevations under study as wilderness. In a landmark case, *Parker v. United States*, local residents and environmental groups argued the agency was prohibited from allowing extraction of timber by the Wilderness Act. The act's provisions provided that primitive areas could not be developed until Congress had a chance to review them as potential wilderness.[3]

Attorneys for the Forest Service, the lumber company, and the Western Wood Products Association argued in part that the road into it had destroyed the area's primitive value. But the court found that the road was inconsequential to the wild character of East Meadow Creek, and it found for the plaintiffs. The case was appealed to the Supreme Court, which upheld the lower court's decision, and in 1976 East Meadow Creek was awarded wilderness status.[4]

Forces were working against the Forest Service's traditional perspective. Multiple-use requirements and the Wilderness Act, along with the Endangered Species Act of 1966 and the Wild and Scenic Rivers Act of 1968, drew public attention to previously unpublicized interactions. Lawsuits brought against the agency for allowing clear-cutting threatened timber operations across the country. The legal disputes led Congress to pass the National Forest Management Act of 1976. It included provisions limiting clear-cuts to forty acres, requiring consideration of biological diversity in timber operations, and stipulating public participation in the planning process.[5]

Use of rangelands was also being scrutinized. A 1972 Forest Service study showed much of the western rangelands were in deteriorating condition. Estimated need for restoration came to $182 million, while fees charged ranchers barely reached $20 million. In 1974 in a federal court case, *Natural Resources Defense Council v. Morton*, the judge ordered the Interior Department to conduct a review of grazing on all public lands. The massive undertaking required 144 land-use plans and thirteen years to conclude. The first completed assessments proposed significant reductions in animal use. Ranchers were incensed. To them, as well as other commodity users, federal regulations were threatening their livelihood. More bad news followed.[6]

The Public Land Law Review Commission, which met from 1965 to 1969, had a directive to consider all western concerns. Based on the commission's findings, in 1976 Congress passed the Federal Land Policy and Management Act (FLPMA). It mainly delineated policies already in place, but, taken together, to resource interests they appeared oppressive.

Among numerous requirements, FLPMA stipulated that lands be

managed for multiple use, that certain of the lands be preserved in their natural condition, and that the United States receive fair value for the lands' use. Most damning of all to some of the largest stock grazers, the act specified that almost all public lands were to be retained in federal ownership.

The Taylor Grazing Act of 1934, in securing ranchers' grazing rights to large areas, had virtually ended the policy of homesteading on federal lands. But it had stipulated that management would be subject to the lands' ultimate disposal. Now, FLPMA made clear that the lands were to be kept by the government in perpetuity. Compounding the rangeland users' grievances, the fair-value policy had already begun to be implemented. Between 1970 and 1976, grazing fees increased from $0.32 per animal unit per month to $1.50.[7]

Jimmy Carter's election in 1977 was, in the eyes of some westerners, the final element of the perfect storm. Early in his tenure, disregarding a drought in the West, he recommended removal of funding for a number of water projects, and he called for further expansion of the wilderness system. He also appointed Cecil Andrus as secretary of the interior. Andrus was a two-term Democratic Idaho governor and a conservationist. A profile of him in the December 1976 *Congressional Quarterly Weekly Reports* said Andrus was "a sure bet to please environmentalists and displease mining, logging and other development interests."[8]

The periodical was accurate in its assessment. Upon taking charge at Interior, Andrus announced he was determined to protect the nation's resources from "rape, ruin, and run." By instituting sweeping changes, he intended "to end the domination of the department by mining, oil, and other special interests."

Environmentalists saw the naming of department officials by Andrus as opening government to people previously excluded. Carl Bagge, president of the National Coal Association, perceived things differently. He said he had begun to pray a lot because "every rock I lift up, I see another professional environmentalist."[9]

In December 1978 Carter took a dramatic step, becoming involved in a seemingly unresolvable land issue. Congress was hamstrung over an Alaskan bill seeking to protect wildlife habitat, including tundra needed by grizzly bears, ancient forests, and alpine lakes. The battle had gone on for four years. The environmental community believed Alaska was America's last chance to protect frontier land rather than "tame" it. But in Alaska commercial mining and timber concerns and local citizens were fighting to keep the land open.

In an effort to settle the issue in Congress, the administration hosted an all-night meeting between Morris Udall, Stewart Udall's younger brother, who would serve as Arizona's U.S. representative for thirty years, and minority whip Alaska senator Ted Stevens. Udall and House Democrats were proposing the withdrawal of 122 million acres from commercial use.

Republican Stevens had forged a deal with Democrat Henry "Scoop" Jackson to create a Senate bill that would protect much of what the House wanted but maintain a variety of opportunities to industrialize. As the night meeting progressed, Stevens threatened to leave the table over one issue or another. Udall would wait for the outburst to end and suggest some cosmetic alternative, and the talks would proceed. By the end of the night, a deal had been struck. Twenty-four hours later, Alaska's junior senator, Mike Gravel, who originally signed off on the proposal, changed his mind and threatened a filibuster that shut down the process.[10]

A moratorium on development was set to expire, and Andrus approached Carter. He recommended that the president proclaim the lands as national monuments. This was the strategy Theodore Roosevelt used to protect the Grand Canyon, invoking the Antiquities Act, in 1908. It could be superseded by congressional action, but until Congress produced a law, the land would be protected. Carter agreed to the move. When Roosevelt had safeguarded 800,000 acres of the canyon, he was criticized for encompassing too large an area. Carter's proclamation protected 56 million acres. The action forced preservation opponents, including oil, gas, and mining interests, and the legislators representing them to continue to negotiate for a congressional solution.[11]

Less than a year into Carter's term, environmentalists were hailing him as the "greatest conservation president of our time" and touting Alaska's untamed image. Alaskans responded, "Image be damned." They protested that their state was no longer America's last frontier; it had become America's last wilderness. One resident commented, "They want food for the soul. We need food for the body." Fairbanks citizens reacted with a violent demonstration, burning Carter in effigy. Two towns vowed to protect individuals who broke the new laws.[12]

The Alaska state government filed a lawsuit, and more than a thousand citizens, reported in some quarters as three thousand, assembled in open rebellion. The Great Denali Trespass took place in January 1979 as protesters converged on Cantwell, a railway flag stop of some two hundred people in the Alaska Range.

The winter was typically bitter, and the demonstrators, carrying "Don't Tread on Me" flags, made their way to makeshift headquarters at the markedly rustic Cantwell Cafe and Bar. They camped out on newly designated monument land, cut firewood, ran dogsleds, hunted, set traps, fired weapons at random, and rode snowmobiles and Arctic Cats, deliberately breaking laws. The National Park Service brought in rangers from the Lower 48, but, sorely outnumbered, they merely watched. No one was arrested, and, the protesters' point being made, they dispersed.[13]

Commodity users in the lower forty-eight states applauded the Alaskan demonstration. Their grievances, like the Alaskans', were complicated by the now well-organized environmentalists. In order to counter them, business interests established opposing organizations.[14]

In 1973 the California Chamber of Commerce created the Pacific Legal Foundation. Advocating free enterprise and limited environmental regulation, it challenged clean-water and endangered species acts' directives that imposed on private property. In 1975 similar goals caused corporate interests to fund the National Legal Center for the Public Interest, ultimately folded into the American Enterprise Institute. Prominent members of the National Legal Center included former solicitor general and future Bill Clinton foe Kenneth Starr and future Supreme Court chief justice John Roberts.

In 1977, after Carter's election, Colorado brewer Joseph Coors and several other businessmen established the Mountain States Legal Defense Fund. Its mission was to promote commercial interests on federal lands. James Watt, the fund's first president, later became Reagan's point man in his administration's efforts to deregulate land management.

In summer 1978 Barry Goldwater and John L. Harmer, a former California lieutenant governor under Ronald Reagan, began to organize the League for the Advancement of States' Equal Rights (LASER). The new group's goal was "to create a broad base of support in favor of divesting the Federal Government of the public domain." It attracted widespread publicity, as joining Goldwater on its board were Utah's senator Orrin Hatch and Alaska's senator Ted Stevens.

In early 1979 a group organized by western ranchers, the Public Lands Council, was encouraging the Nevada Legislature to pursue state control of federal lands. Rancher and Nevada assemblyman Dean Rhoads was the council's president. In 1975 Nevada had funded a study that endorsed state procurement of the lands. The study and Rhoads's group's lobbying had their effect. On February 16, 1979, Nevada's *Elko Daily Free Press* announced: "First shots fired in sagebrush rebellion."[15]

The newspaper was referring to the state assembly's introduction of AB 413, declaring state sovereignty over the BLM domain, 68 percent of its land. The bill would provide a war chest of $250,000 to test the new law in a court case. Rhoads had garnered thirty-seven of forty assembly members as cosponsors of the bill, and the senate measure had fifteen of twenty members as cosponsors. Quotable state senator Norm Glaser, an Elko rancher, later used Old West idioms to explain the state's action. Federal agencies, he asserted, "dry gulched the cowboy, bushwhacked the miner, and ambushed the sheepherder." In order to get the matter to court, he announced, "We're going to have a head on confrontation. We're going to arrest all the BLM."[16]

Two months later, some five hundred Nevadans amassed at a legislative hearing to discuss the bill, "likening their actions to the attack on Fort Sumter." New BLM policies were characterized as a "great federal land grab." Letters of support for the legislature came from both Nevada U.S. senators and Governor Robert List, who looked past state ownership to urge privatization. He wrote, "I endorse getting as much of the productive land as feasible into private ownership."

At the hearing one senator, Clifton Young—later a justice on the Nevada Supreme Court—spoke against the bills. He said that faces glowered throughout the big hearing room when he started to speak. "I had sort of a tongue-in-cheek approach," he said. "I was kidding them and pointing out how ridiculous it was; that the worst thing in the world that could happen to us is if they won the right to have all that federal land. It would bankrupt the state! [Higher fees] would drive the [ranchers] out of business." He argued that the state would have to sell the land to the highest bidder, creating poorer land and watershed quality and denying recreational opportunities. As Young remembered it, an old lady rose, pointed a crutch at him, and said, "Either let's shoot the son of a bitch, or we ought to hang the son of a bitch." News reports merely said that Young's speech was answered by "a loud chorus of hisses and boos."[17]

Although beginning with a publicity bang, there was no blueprint for how the Nevada rebellion would proceed. In early June the *Nevada State Journal* published an article titled "Nevada, Rebel without a Plan." It pointed out that the cause had been identified but the instruments for action had not. Two months on, when asked by a BLM official what the plans were for implementing the law, the state registrar responded, "That's a question I can't answer."[18]

By July Nevada had formulated its new regulations. Land-use applicants were required to get written authorization from the registrar while still having to obtain leases and permits from the BLM. Once approved for land use, the applicant needed to provide information to the state regarding fees, rents, and royalties paid to the federal government. Another couple of layers had been added to the despised bureaucracy.[19]

Despite the practical difficulties, four other western state governments began to pursue legislation similar to Nevada's. In 1979 and 1980 New Mexico, Utah, and Arizona—which had to override the veto of Governor Bruce Babbitt—passed legislation similar to their neighbor; Wyoming went further, claiming Forest Service land as well as BLM land. Hawaii passed a resolution of support for the rebellion. Four other states, California and Colorado (each of whose Democratic governors vetoed rebellion laws passed by their legislatures), Idaho (which drafted at least four separate bills), and Alaska, adopted resolutions for feasibility studies to pursue land conveyance. Washington State's legislators tied a sagebrush bill to a constitutional amendment eliminating a clause that kept unappropriated land in federal ownership, but that proposition failed with a 60 percent "no" vote. In 1979 bills were considered in the Oregon Legislature, but they died in committee. The Montana Legislature met later and, owing to the lobbying of a coalition of environmentalists, hunters, fishermen, and other outdoor enthusiasts, also defeated proposed legislation.[20]

At the time, political leaders were issuing provocative statements. Senator Hatch of Utah was saying the struggle was "the second American Revolution." Alaska's lieutenant governor, Terry Miller, proclaimed the West a victim at the mercy of "ignorant and arrogant husbandry by a distant and often uncomprehending federal government." He claimed officials' actions had redrawn the Mason-Dixon Line to run north and south and separate the East from "an increasingly angry West."[21]

Environmentalists characterized the rebellion as a land grab by large ranch owners and speculators. They echoed Clifton Young's pronouncement about Nevada, saying that western states had not done well in managing state-owned lands. They pointed out that Nevada had sold practically all the land granted it upon admission to the Union, retaining only 1 percent. States that maintained larger percentages of their land, on a per-acre basis, allotted fewer moneys and assigned fewer administrative staff than federal properties. An extreme example was Colorado, where four field appraisers were responsible for managing more than three million acres.[22]

The *Los Angeles Times* weighed in, saying it was not sure whether the rebels' leaders were patriots or mercenaries. It concluded that the paper never backed down from a fight with the federal government, "but the notion of turning over territory held in trust for all Americans to the capricious management of the states strikes us as a dubious cause."[23]

The Interior Department's top lawyer, saying there had to be a better way to handle the states' complaints, said, "I wouldn't give a plugged nickel for litigation that tries to take land from the federal government." He said, for the time being, the department intended to ignore the Nevada law.[24]

Interior Secretary Andrus called the Sagebrush Rebellion the "great terrain robbery." He issued a lengthy press release, saying the movement was an attempt to "hornswoggle" Americans out of their unique land heritage. Calling to mind Bernard DeVoto, the secretary commented that the land some state officials proposed confiscating "belongs to all the people of this country" and is "a legacy of freedom and openness for our children and grandchildren." He argued that states would not be able to manage the lands unless they raised taxes. Because few would be willing to do that, states would be forced to sell the property piece by piece.[25]

The Andrus news release prompted Nevada's U.S. senator Paul Laxalt to respond. Laxalt sent the secretary an open letter. "Westerners are not known as hornswogglers," he wrote. "They are known for their lack of hypocrisy, their open, straightforward approach to life and their willingness to live and let live." The senator argued that some federal bureaucrats felt the need "to inject themselves into the daily lives of public land users," stripping the West of its ability to control its own destiny.[26]

In September 1979 California governor Jerry Brown vetoed the bill to enlist the state in the Sagebrush Rebellion, commenting, "As stewards for the future, we should direct our attention to improving land management as we debate which jurisdiction should legally control it."²⁷ Brown's comment, accepted, however grudgingly, in Sacramento, would not have served him well in certain other western locales.

Moab, in Grand County, Utah, is a desert town in red-rock country that features spectacular buttes, mesas, arches, spires, and gorges. Originally settled by Mormon cattlemen, it now hosts mountain bikers, four-wheel-drive enthusiasts, hikers, river runners, and rock climbers who utilize the nearby Arches and Canyonlands National Parks. In 1979 the county population was roughly eight thousand and the town's perhaps fifty-three hundred. At that time, roughly one-third of the labor force in Grand County was employed in agriculture or mining. In Moab an escalation of the Sagebrush Rebellion began with intensifying angry rhetoric directed at the BLM. Senator Hatch said the BLM men were "stumbling over each other, acting like little gods." He alleged: "They're being paid for nothing but to cause trouble." At a BLM meeting a county commissioner threatened open rebellion: "I'm getting to the point where I'll blow up bridges, ruins, and vehicles. We're going to start a revolution."

Hardscrabble wilderness philosopher Edward Abbey had portrayed the area in his provocative novel *The Monkey Wrench Gang*. The story takes place mainly in San Juan County, adjacent to Moab. The activist heroes of the book use vandalism to sabotage developments that have damaged the environment. The nemesis of the protagonists, a developer named Bishop Love, was based on San Juan County commissioner Calvin Black.

To those who knew him, Black was magnanimous and good-natured, although it was said that having an argument with him was like "riding a bull." Black called himself one of the "true environmentalists," but he made no apologies for his career trying to maximize economic growth and limit government.[28]

In 1979 Black attended a BLM meeting and was quoted during a tirade directed at a BLM employee: "We're going to get our lands back. We're going to sabotage your vehicles. You had better start going out in twos and threes because we're going to take care of you BLMers." Interviewed later by the *Deseret News*, Black said he was not issuing threats, that he was only repeating what he was hearing people say. But the minutes from the meeting record a question from the BLM official asking if Black was threatening him, and the commissioner responding, "I'm not threatening you, I'm promising you." A Utah bumper sticker read: "BLM—Better Leave Moab."[29]

7

Harnessing Forces

In 1980 the Sagebrush Rebellion was gaining momentum. The rebels appeared to be building the foundation for a widespread movement. In retrospect, cracks were already developing.

A year after Nevada claimed federal lands, state officials still had not decided how to mount their legal challenge. The western states' attorneys general were scheduled to meet in mid-July, when strategies could be discussed. But none of the state legislatures had appropriated funding for the cause. Nevada's attorney general was questioning the lawmakers' plans. He commented, "If the [Nevada] legislature has had a change of heart on this thing, my point to them is, don't leave me hanging out there." The state's legislators were in fact having second thoughts. The senate's finance chair said, "It's just a case of if you can't win it, I don't think we ought to go after it." Other states' officials had also balked. An analysis by the Alaska Legislative Affairs Agency concluded, "In our opinion, and in the opinion of other attorneys who have examined the equal footing doctrine, there is little or no hope that it will stand up to judicial scrutiny."[1]

Still, with numerous entities in the western states in seeming agreement with the concept of a rebellion, avenues other than court challenges appeared promising. As the rebellion commenced, Utah's Hatch had produced a Western Lands bill in the U.S. Senate. Nevada representative Jim Santini patterned a House bill after it, and, with revisions in mid-July 1980, twenty cosponsors signed on. The proposed bills, which would provide for the cession of federal unreserved and unappropriated lands to the states, had bipartisan supporters but had not moved out of their respective committees. As the presidential campaign took shape, Ronald Reagan, influenced by his national campaign chair—Nevada senator Paul Laxalt—announced his support for the rebellion. The yet-to-surface problems arose from the lack of consensus on the crusade's basic goal: Should public lands be conveyed to the states or privatized?[2]

At the same time, in 1980, there was an anomaly: an instance of the interests of the rebellion advocates and environmentalists coalescing. The Santini-Burton bill, in the House of Representatives, involved selling BLM lands near Nevada's Las Vegas Strip. The sales would procure funding for restoring damaged lands and purchasing environmentally sensitive properties at Lake Tahoe.

Lake Tahoe is a spectacular body of water 22 miles long, 12 miles wide, and 1,636 feet deep—the third-deepest lake in North America. It sits atop the northern Sierra Nevada, dissected by the California-Nevada state line. The lake was home to the Washoe Indians until the gold and silver rush in the second half of the nineteenth century, when Euro-Americans overwhelmed them, pushing them out of the lake basin. In the 1870s the forests on the precipitous mountains surrounding the lake were clear-cut to provide lumber for securing the walls of Nevada's deep Comstock Lode silver mines. Removing the basin's vegetation caused erosion that clouded Tahoe's previously pristine water. Over the course of decades, the forests grew back and the lake regained its transparency.

But beginning in the 1960s, with the area's beauty reestablished, Tahoe became a tourist haven. Removing hillside vegetation for buildings and roads again caused erosion impairing water clarity. And battles erupted between those trying to protect Tahoe's ecology and developers and boosters.[3]

In 1979 there was some agreement that the government could protect the lake from further damage by purchasing environmentally sensitive lands in the basin. The League to Save Lake Tahoe first floated a plan that Nevada representative Santini began to champion. The idea was to sell BLM lands around Las Vegas and use the money to buy properties at the lake. In July 1979 Santini sought assistance from California's powerful U.S. representative Phillip Burton.[4]

Democrat Burton was the chair of the House National Parks Subcommittee. First elected in 1964, he was a fire-breathing preservationist with substantial influence. He had steered the Redwood National Park expansion through Congress and authored the bill creating the Golden Gate National Recreation Area. The Point Reyes National Seashore includes the Phillip Burton Wilderness. Burton was intrigued by Santini's idea, but he strongly opposed a massive sell-off of federal lands, insisting on only a small-scale experiment.[5]

The two legislators met at night for months, arguing over what a bill might look like. Santini, in support of the Sagebrush Rebellion, wanted a large sell-off of properties to jump-start the conveyance of other federal lands, while Burton vehemently advocated a limited sale. Regarding Burton, Santini commented: "His mind worked so fast that it was a challenge to keep up." And Burton complimented Santini: "Goddamn it Jim, one thing I like about you is that you're operational. I hate those fucking ideologues." The representatives worked into 1980. Fearing further construction at the lake, Burton wanted zoning laws or land-use controls written into the bill; Santini, who represented owners of exclusive Nevada properties and the large Tahoe hotel-casinos, objected.[6]

The obstacles seemed overwhelming. Not only did Santini have to deal with powerful economic interests, but he also had to convince the people of Clark County, where Las Vegas is located, that selling lands in their county to help resolve environmental problems elsewhere was a good idea. On Burton's side, environmentalists were leery that the scheme might lead to a wholesale divestiture of federal lands. No one felt inclined to trust anyone.

In May 1980, after thirty-six closed-door meetings, the representatives introduced the Santini-Burton bill, authorizing the sale of up to nineteen thousand acres of BLM lands in Clark County. The Forest Service would have up to $150 million to buy sensitive lands at Tahoe. By the end of the year, both houses passed the bill, and President Carter signed it into law.[7]

While the government was selling public land in Nevada for a specific environmental purpose, in Moab local leaders emphatically registered their differences with federal regulatory policy. For months incendiary rhetoric had marked discussions in southeastern Utah. County officials perceived federal decision makers to be unresponsive. To demonstrate their support for local agriculture and mining, they decided to rebel openly.

In 1866 Congress had passed Revised Statute (RS) 2477 in its mining laws, which contained a sentence that would create controversy far into the future. The statute granted "the right-of-way for the construction of highways across public lands not otherwise reserved for public purposes." A highway referred to any road or right-of-way. Although Congress repealed the highway provision in 1976, existing claims were grandfathered in.

What constitutes a road became critical in the mid-twentieth century when an important part of the notion of wilderness was tied to an area being roadless. Congress provided that if a roadway built by means beyond just vehicle travel had been in continuous use before 1976, a county can assume ownership. The BLM ultimately separated a "way" from a road. It defined a way as a temporary passage formed only by vehicle travel that did not impinge on an area's wilderness setting. County officials in Utah disagreed. Insisting roadways and tracks are vital parts of their heritage and might still be used, they have claimed some twenty thousand "roads."[8]

Such a track became the focal point of Moab's rebellion on July 4, 1980. The town held a rally of some three hundred people at the edge of a BLM Wilderness Study Area. The local chapter of the American Legion presented the colors as the national anthem played, and politicians gave speeches from the back of a Caterpillar bulldozer. An American flag flew from the D9 Cat's smokestack. Bumper stickers affixed to the machine read "SAGEBRUSH REBEL."[9]

At the 1980 Moab rally, speechmaking by several politicians included references to the "cancerous" growth of bureaucracy. Speakers contended that because the new federal land policies were "devastating to our country," southeastern Utah would take control of its own destiny. When the last speaker stepped down, county commissioner Ray Tibbits manned the bulldozer. Following traces of an abandoned mining road, he scraped a path toward the study area. The rebels had misread the map, and Tibbits stopped a quarter mile short of his destination. The following Monday, when apprised of the situation, the commissioners ordered a county grader out to complete the act of civil disobedience.[10]

The U.S. attorney for Utah informed the county commission that the bulldozed area needed to be restored, or it would be done by the BLM and the county would be billed. He went on to declare that laws would be upheld and that agencies, including his office, would ensure U.S. lands were protected. The commissioners capitulated and restored the land.

A month later John L. Harmer, cofounder and executive officer of LASER, the League for the Advancement of States' Equal Rights, spoke about the Moab confrontation. He called it a classic example of "abuse and oppression." He claimed it demonstrated "the federal government's contempt for local authority and interests, and their determination to eliminate any vestige of local control over the public domain."[11]

Shortly thereafter, presidential candidate Reagan, who abhorred government interference with business, announced he was to be counted in as a Sagebrush Rebel. At the Republican National Convention in 1980, Reagan declared, "We will not permit the safety of our people or our environmental heritage to be jeopardized, but we are going to reaffirm that the economic prosperity of our people is a fundamental part of our environment."[12]

For years, in the 1950s, Reagan had represented General Electric, traveling to GE plants, personifying what the company hoped would be seen as its benevolence. His speeches at the plants asserted the benefits of unfettered capitalism. GE executive Lemuel Boulware, who spent considerable time and money recruiting business executives to become involved in politics, is widely acknowledged as having had a profound influence on Reagan's conversion from fervent FDR supporter and "Labor for Truman" organizer to the champion of modern conservatism. As Reagan's political perspective changed, he reversed the roles of the common man and corporations, identifying big business as being victimized.[13]

Nevada's Laxalt, Reagan's close friend since the 1960s when they were governors of adjoining states, commented, "[Reagan] had a fundamental philosophy, which evolved during the General Electric days when he went all around the country as a spokesman for them. He was a vehicle, their vehicle, for private enterprise, extolling the benefits of private enterprise and all the evils of excessive government. He had had a very simple philosophy that he had developed, and that made it very easy for him wherever he went."[14]

Reagan's radio show in the late 1970s provides insight into his attitude toward the environment in the years leading up to his presidential campaign. Included among a large number of similar subjects: in July 1978 his topic was titled "Allow Cutting in National Forests"; in October 1978, "Make Public Resources Available to Industry" and, later that month, "Produce More Energy from Federal Lands"; in February 1979, incredibly, "80% of Air Pollution Comes from Plants and Trees"; and in September 1979, "No Alaska Lands Bill."[15]

While campaigning Reagan consistently attacked the Clean Air Act. He decried excessive regulations, promising to fight "hyper environmental groups" and EPA bureaucrats, whom he accused of "trying to create deliberate traffic jams to keep people from driving cars." A plank of the Republican platform pronounced that "environmental protection must not become a cover for a 'no growth' policy and a shrinking economy." And Reagan disparaged environmentalists, saying they were "protecting rabbits' holes and birds' nests."[16]

Nixon's Watergate fiasco, and his subsequent pardon by Gerald Ford, had taken its toll on the Republican Party. In 1977 a Gallup Poll had shown that only 20 percent of U.S. voters supported the party, its worst performance in forty years. By 1980 Reagan was appealing to disaffected southerners and westerners. But as the campaign progressed, Reagan's managers began to fear that his antienvironmental remarks were alienating too many voters. They formed an advisory committee of prominent Republican environmentalists. Russell Train and William Ruckelshaus, who each had previously headed the EPA, were on the committee, which provided the campaign with a more positive environmental message.

Then in 1980 international and domestic affairs seemed to spin out of control, casting a dark shadow over the Democratic administration. The seemingly interminable Iran hostage crisis was a staple of news reports. In April an attempted rescue of the hostages turned into a disaster, with helicopters failing and one crashing into a rescue airplane, killing eight and injuring five. More than one hundred thousand Cuban refugees arrived in America—a group of whom, it turned out, were criminals, released as Castro emptied his prisons.

A summerlong heat wave, amid the worst drought since the Dust Bowl in the South and Southwest, caused more than twelve hundred deaths. With the acquittal of four policemen who fatally beat a black man, Miami riots caused fourteen deaths, three hundred injuries, and one hundred million dollars in damage. Abscam, an FBI sting operation, uncovered thirty-one public officials willing to take bribes from an agent posing as an

Arab sheikh. Overarching all other ills, the economy was faltering. Unemployment was up, and inflation rose to 12.4 percent, the second year in a row in double digits.

The Democrats' 1980 strategy, like the 1964 campaign, was to portray the Republican, talking tough about foreign affairs, as a dangerous figure who might spark a war. But rather than Goldwater's prickly demeanor, Reagan, at sixty-nine years old, had a winsome manner and vast experience in the public forum. He seemed optimistic. Laxalt later described Reagan, saying, "He spoke about the same stuff for 15 years after I first heard him in 1963, and he would hardly change a phrase. He fundamentally believed that stuff, and the people finally accepted it. He may be an actor, and he may not be an intellectual. He may not be overly energetic. But, damn it, he believes. Whether you agree with him or not, we know he is sincere."[17]

Reagan got on the stage with President Carter for only one debate. It was all he needed. The defining moment came as Carter brought up the fact that Reagan had campaigned against Medicare throughout the 1960s. Reagan did not answer the charge. He replied simply, "There you go again." For many Americans, the response was an effective answer, and Reagan went on to a landslide victory.

Winning forty-four states, the Republicans also defeated twelve incumbent Democratic senators to take control of the upper house. But the results were widely misinterpreted. Early in the 1970s the public believed social issues and economic issues to each be important problems in the country. By 1979 70 percent of Americans said economic problems were most important; only 10 percent identified social issues as critical. The Republican Party touted the themes of increasing conservatism and anger at government, and they became important parts of Republican ideology thereafter. But election polls and studies showed that the voting had been partly, or largely, a President Carter plebiscite. Nearly 85 percent of new Republican voters disapproved of Carter's performance, and 91 percent disapproved of his handling of inflation. Their responses showed them to be neither especially conservative nor angry at the government.[18]

In November, immediately after Reagan and the Republicans won, the first and only LASER conference was held in Salt Lake City. More than five hundred attendees gathered to hear how to go about "divesting the federal government of the public domain."

At the conference Nevada's Dean Rhoads explained that with the current Supreme Court configuration, it seemed any Sagebrush issue would be a five-to-four ruling against the movement. There was much better news when rebels considered the other branches of government. The proposals by Hatch and Santini, with the Republican congressional gains moving western senators into key leadership positions, became decidedly more promising, and President-Elect Reagan telegraphed the LASER conference, saying, "I renew my pledge to work toward a Sagebrush solution."[19]

Senator Hatch was a keynote speaker at the conference. He exhibited his rebel bona fides, calling environmentalists "dandelion worshippers" and those in government who sustained environmental issues "land embalming park managers."[20] Hatch told the LASER audience that the establishment of a strategy reflecting a mutual goal was needed. "Only with a unified effort," he said, "can we expect to meet the challenge before us." But dwarfed by the enthusiasm the landslide election generated, it went largely unrecognized that different attendees expressed different, potentially divisive, goals.

While significant numbers at the conference wanted lands transferred to the states, LASER organizer John Baden and Alaska senator Ted Stevens spoke of "movement toward the private sectors." Former BLM director Marion Clawson favored innovation, but did not solely support either transference or privatization. Rhoads said the rebellion's direction lay in piecemeal legislation, executive action, and adjusted regulations.[21]

Reagan's close associate Laxalt further jumbled the message, promising that Reagan would push to give states control of more than 700 million acres. But upon his return to Washington, he said the Sagebrush Rebellion was actually only a "symbolic thing." He explained, "When we get people in key positions, who recognize the West as an equal partner, that will take care of a lot of our problems."[22]

In late November 1980, before giving up his office, Interior Secretary Andrus began one last push to resolve the Alaska land issue. Morris Udall and his allies realized that the incoming new majority in the Senate would not provide the environmental protections for which national conservationists had worked. The current Senate bill was full of compromises that the Democratic House had repeatedly rejected. Now, in the lame-duck session, the House had to either capitulate or face dwindling chances of protections getting approved.

Initially, some hostile conservation groups argued that Reagan might get them a better deal. With Andrus calling their argument idiotic, other environmentalists quickly overrode the obstructionists, helping push through compromise legislation. The bill set aside 103 million acres, 56.7 percent of which was designated as wilderness; the rest received various levels of protection. The new wilderness designation brought America's wilderness land to 100 million acres. Andrus commented that the process proved the old adage, "There's nothing like a hanging in the morning to focus the mind." When Jimmy Carter left office, he called the Alaska act his single most important legacy.[23]

Andrus had one more important action he hoped to complete before January 20, 1981, when the Reagan administration would take over. He wanted to give permanent protection to five rivers in Northern California. His action had been challenged in court, and Andrus had to wait until the U.S. Ninth Circuit Court of Appeals upheld his authority to designate them wild and scenic. The court did so late on January 19, and, on his last night in power, he drove to his office and signed papers that protected the American, Eel, Klamath, Smith, and Trinity Rivers. Reagan's incoming interior secretary, James Watt, threatened to cancel the directive, claiming that Andrus was no longer the secretary. But by law Andrus was able to carry out official duties until the moment Reagan took the oath of office, at noon on January 20.[24]

Other groups, seeking to use the rivers for Southern California development and farming, raised other objections. In January 1985, at about the time Reagan was taking the oath of office for his second term, the announcement was made that the Supreme Court was letting the judgment stand, allowing the protection of the rivers.[25]

It was another win for environmentalists. To extractive commodity interests, the established decision-making process had lost all balance. But, as indicated by Watt's intercession, a radical adjustment in federal influence was imminent.

8

Fracturing Policies

"In the present crisis, government is not the solution to our problems; government is the problem." In his first inaugural address, Ronald Reagan was speaking about economic ills. But the quote came to be a Republican Party truism cited in a variety of matters, including land issues. Reagan's statement heartened the Sagebrush Rebels. In disparaging national solutions, it seemed plausible that the president would assign management of lands to the states.[1]

When he arrived in Washington, Reagan received two transition taskforce reports on the environment. The first, from members of previous Republican administrations, proposed that the impetus of environmental safeguards be maintained while restrictions were eased. He set aside that recommendation for one by the Heritage Foundation, a conservative think tank funded by beer magnate Joseph Coors and billionaire banking heir Richard Mellon Scaife. The Heritage Foundation document called for massive changes on public lands. At an early cabinet meeting, Reagan distributed the report to each officer, directing them to implement the items relevant to their departments.

Reagan's goal was to develop the country's resources and reduce spending. The Carter administration had projected an increase in spending on the environment from thirteen billion dollars to sixteen billion in 1984. The Reagan administration planned to reduce the amount to nine billion dollars. Reagan officials, promoting natural resource extraction as a benefit to national security, portrayed environmentalism as antithetical to American ideals. But the president's widespread popularity did not extend to his environmental policies.[2]

Reagan's choice for secretary of the interior, James Watt, was especially controversial. He had been the first director of the Mountain States Legal Foundation, which entered into forty-seven lawsuits challenging federal regulations. When it became known that Watt was to be named, twelve major conservation groups sent a telegram urging reconsideration. David Brower commented that if Watt was confirmed, "the war against the earth would step up in pace, and this country and the globe are not likely to reach the year 2000 intact."[3]

On January 7, 1981, Watt's confirmation hearing room was packed with spectators and seventeen television cameras. Watt, six-foot-five and rail thin, was composed and well spoken. Although saying he believed there was excessive bureaucratic regulation, he pledged that if confirmed, he would use a commonsense approach to balance managing land and water.[4]

The following month, at a hearing before a House committee, Watt created a nightmare for himself. The questioning was cordial, and he again presented himself well, but in one reply he said too much. Asked if he believed that some resources should be left for our children, Watt said he absolutely agreed. He then qualified his answer: "I do not know how many future generations we can count on before the Lord returns. Whatever it is we have to manage with a skill to leave the resources needed for future generations."

Watt's remark, reflecting his born-again religious bent, became fodder for political commentators. It was brought to the public's attention in June 1981 when Herbert L. Block, *Washington Post* award-winning cartoonist, produced a political-page cartoon titled *Onward, Christian Soldier*. Block showed Watt marching ahead of a bulldozer labeled "Timber and Mining Interests." He was leading it into forested mountains identified as "U.S. Lands and Resources." A jubilant, cigar-smoking businessman was at the wheel of the bulldozer. A worried Uncle Sam stood nearby, and Watt carried a sign saying, "Why Save it? THE END IS NEAR."[5]

Watt was the easiest of Reagan's appointees to characterize as antienvironment because he was the most outspoken. Others were equally ideological. Reagan's intention to emphasize resource production became apparent in his appointments.

For assistant secretary of agriculture in charge of the Forest Service, he appointed John Crowell, general counsel of the timber giant Louisiana-Pacific. Robert Harris, the organizer of litigation against the surface mining act, was tapped to head the office that oversaw surface mining. Robert Burford, a Colorado rancher, state legislator, and self-proclaimed Sagebrush Rebel, was named director of the BLM. Burford's future wife, Anne Gorsuch, a James Watt associate, served as the most controversial administrator in the history of the EPA.[6]

As Reagan put together his team, the Heritage Foundation had provided a "hit list" of unacceptable scientists for the EPA. Whether it was used was disputed, but it was in the possession of a Reagan official, illustrating the extent of the administration's resolve to change environmental policy. The list comprised the names of ninety scientists and included comments next to names such as: "Get him out, horrible," "Clean air extremist," and "reported to be a liberal and environmentalist."[7]

A major theory involving policymaking is a hypothesis concerning stakeholders' perceptions in policy conflicts called "the devil shift." Developed by Paul Sabatier, Susan Hunter, and Susan McLaughlin in the late 1980s, the theory deals with the concept that those with opposing views are seen as stronger and more evil than they actually are. The actors will impugn the motives of their adversaries while seeing themselves as reasonable. And they will use incendiary rhetoric or media onslaughts to dehumanize the opposition and portray the struggle as one of "good versus evil." James Watt was consistently depicted as promoting clear-cutting the forests. False accusations of statements included "After the last tree is felled, Christ will come back."[8]

On the other hand, in a book in support of Reagan as a Sagebrush Rebel, William Pendley, Watt's deputy assistant at Interior, described the president's environmental foes. In the book's introduction the author labeled them variously as "radical," "childish," and "arrogant" as well as "tyrants" and "extremists" exhibiting "rage" and "inflamed passions." In one sentence he listed entities abetting them: "mischievous laws," "covetous bureaucrats," "a feckless Congress," "duplicitous media," and "a lavishly funded environmental lobby."[9]

In pursuing development in previously protected areas, the Reagan administration listed the possibility of Russia and other nations limiting minerals essential to America's industry and defense. Fears of Middle East oil embargoes were the stated impetus for expanding coal, oil, and gas exploration; surface mining; and offshore drilling. Watt's plan for waters along the nation's coasts included drilling across two hundred million acres a year, for five years, fifty times the amount of acreage opened in U.S. history.[10]

Denver's *Rocky Mountain News* interviewed Watt on December 23, 1980, the day after his nomination to become secretary of the interior. Although previously supporting the Sagebrush Rebellion, he had come to the conclusion that the conveyance of massive amounts of federal land would be a "waste of money." Instead, he admitted that the BLM had a right to decide how federal lands should be used. His goal at Interior, he said, would be to change attitudes, to make sure that the lands were managed as "a good neighbor." This would "let the Sagebrush Rebellion die because of friendly relations."[11]

Immediately after assuming his post, Watt opened communications with western state officials. Advocating a "commonsense" approach, he began supporting land users' know-how over scientific analysis and planning. He shifted federal powers to the states, allowing 99 percent of the country's working coal mines to be regulated by state governments. Other major changes Watt instituted included easing surface-mining regulations; dramatically increasing amounts of timber to be cut in Oregon and California; putting a moratorium on federal purchases of lands, insisting the government could not manage what it had; proposing mineral leases within existing wilderness areas; and dismantling off-road-vehicle regulations.[12]

Environmental leaders insisted Watt was exceeding his authority. He answered their criticism by rejecting them as legitimate participants in government affairs. He met with them in early 1981. At the conclusion of the talk, he announced there would be no more such meetings. "[Environmentalists] were quite surprised and upset when we did not consult them on decisions," Watt later commented. "But we didn't need to. We knew exactly what we wanted to do."[13]

The Republican Party wholeheartedly supported an oil policy later famously touted as "drill, baby, drill." But where to drill could be subjective.

In August 1981, after several years in which "untold man hours" were spent preparing it, a voluminous document proposed oil and natural gas exploration just south of Jackson Hole, Wyoming.

Watt's deputy Pendley carried the proposal to a Wyoming congressional delegation: Republican senator Malcolm Wallop and representatives of Republican senator Alan K. Simpson and future vice president Dick Cheney—who owned a part-time residence outside Jackson Hole. When the positive recommendation was presented, Wallop declared he had already heard from oilmen in Casper who opposed it. Pendley asked what the oilmen had thought of a similar proposal in the Bob Marshall Wilderness area the previous year. Some of the staffers smiled or smirked as Wallop responded, "They don't have summer homes in the Bob Marshall."

When President Reagan asked Watt why he was going to deny the Jackson Hole project, Watt replied, "Three reasons: Wallop, Simpson, and Cheney." Reagan responded "Jim, if you do not do it, who will? If not there, where will we drill?" The response emboldened Watt, but the Jackson Hole drilling was not undertaken, as the "not in my backyard" politicians were joined by the State of Wyoming and environmental groups, and their lawsuit blocked the project for the duration of Reagan's tenure.[14]

Enough of Watt's other proposals went forward so that, in October 1981, conservation organizations carried a petition to Congress, calling for Watt's removal. The document had 1.1 million signatures, the largest citizen petition ever brought to the Capitol. The same month, at a membership meeting in St. Louis, the Humane Society proclaimed that its 120,000-member organization believed Watt should be removed in favor of a secretary who believed in stewardship and conservation of the country's wildlife and lands.[15]

Reagan ignored the criticism. By providing producers with expanded rights, he had radically changed the balance of land policy. Both the advocates of states' primacy and those interested in privatizing the public domain were heartened by the action, but a rift was developing.

The good-neighbor policy disappointed the Sagebrush Rebels, whose goal was conveyance of land. LASER was continuing its push to get lands into state hands, although the organization was faltering from lack of funds. Hatch and Santini had reintroduced their transference bills, modified to include assurances that almost all new state lands would be kept from private buyers.

The rebels' original plan to battle over the public domain in the courts had been discarded. The rebels had argued that because eastern states had substantially smaller areas under federal management, the equal-footing doctrine had not been properly applied. Their assertion, rebutted by legal precedence from the *Stearns v. Minnesota* Supreme Court decision in 1900, had been further clarified in 1949. In *United States v. Texas*, Justice William O. Douglas wrote, "Area, location, geology and latitude have created great diversity in the economic aspects of the several States. The requirement of equal footing was designed not to wipe out those diversities, but to create parity as respects political standing and sovereignty." The two decisions made clear constitutional parity referred to states' political standing, not property.[16]

With the legal issue blocked and with no success in attempts to move the Hatch and Santini bills in Congress, the efforts of Reagan's executive branch became all important. A key part of the Reagan strategy in promoting development involved the use of the Office of Management and Budget (OMB). It had been designed primarily to shape budget presentations to Congress. But owing to its unique position as an office unregulated by Congress or the courts, it had been an effective tool of executive authority in the Nixon and Carter administrations. Reagan expanded its powers of approval over regulatory proposals. Funding for information services for all agencies was cut, as were appropriations for citizen participation in agency matters such as water-quality planning.[17]

The EPA was a primary target. Reagan placed it under direct OMB control and severely cut its budget. Anne Gorsuch had a twenty-two-month tenure as the director of the EPA. She wore furs, smoked two packs of Marlboros a day, and drove a government-issued gas-guzzler that got fifteen miles per gallon. She was a friend of James Watt, perceived by some to be his "clone or a protégée," and she strongly supported Reagan's goals regarding the environment.

Gorsuch declared that the EPA was too big, too wasteful, and too restrictive on business. Appointees who worked under her came almost entirely from the industries they were charged with overseeing, and both research and enforcement of regulations were significantly reduced during her reign. Republican environmental adviser Russell Train believed Gorsuch was attempting to destroy the agency and deplored the "demoralization and institutional paralysis" she created.[18]

From her arrival Gorsuch confronted the agency's civil service personnel. An administration official said, "She didn't handle things with much finesse starting off, in part because she wanted to show that she was a real soldier."[19] She was accused of mismanagement of the $1.6 billion program to clean up hazardous-waste sites. Acting on the advice of Reagan and the Justice Department lawyers, she refused to turn over agency records to Congress. Charged with contempt, in March 1983, she resigned, saying later that she was "jettisoned" for following orders. Her legal problems were compounded when the Justice Department refused to represent her because they were investigating charges of EPA corruption.[20]

As the Reagan presidency progressed, Gorsuch's ally Watt entangled himself in successively more difficult political predicaments. He was continually fighting Congress in attempts to loosen laws regarding strip mining, offshore oil drilling, and coal leases. Regarding coal, his stated goal was to sell more before 1985 than had been sold in all of U.S. history. The coal industry was mining 100 million tons a year. In April 1982 he sold 1.7 billion tons along the Powder River in Montana and Wyoming, the largest sale ever. Bids were low and favored certain companies. Much of the sales went to AMAX, Shell Oil, and Montana Power. William Coldiron, Watt's solicitor at Interior, was a former board member for Montana Power, and all three companies were contributors to Watt's previous employer Mountain States Legal Foundation. The Government Accounting Office found that remunerations from the sale of the reserves were $100 million less than what should have been generated.

The following year Watt challenged the legislature's authority. He was determined to sell 550 million tons of coal in an auction at Billings, Montana. He was warned by the House Technology Assessment Office, the General Accounting Office, and the House Interior Committee staff that it was a bad idea because of the glut of coal on the market. On August 3, 1983, Democrat Udall's House Interior Committee, with oversight jurisdiction, voted twenty-seven to fourteen, along party lines, ordering Watt to delay the sale. They were still looking into whether the disastrous Powder River bids had been competitive.

With solicitor Coldiron concluding that the House lacked authority to delay the Billings auction, it went on as planned. Only four small companies took part, and prices were less than a penny a ton. The White House squirmed as environmentalists brought lawsuits. Congress fumed, eventually placing a moratorium on coal leases and curbing offshore oil and gas exploration.[21]

Now the divergent principles of states' rightists and advocates of privatization came to the fore. Watt's good-neighbor policy had subverted the Sagebrush Rebellion by discounting divesting lands to the states. A group of Reagan advisers had subsequently begun arguing that lands should be sold outright as a national debt-reduction measure.

Reagan disliked bureaucratic conflict. He allowed aides and cabinet officials to pursue their individual objectives, and he would ratify whatever compromise evolved. But now the time had come for him to decide the difficult policy question of possible disposition of federal lands to state or private interests. In his February 1982 budget message, Reagan announced that some government land would be of greater value to society if it was privatized. He pledged to "move systematically to reduce the vast federal holdings," not by transference but by sales.[22]

Steve Hanke, on the president's Council of Economic Advisers, couched the "Federal Real Property Initiative" as "a choice between capitalism and socialism." He argued that sales would reduce the federal government, benefit commodity users, and create a state and local tax base.[23]

The proposal caused the already weakened Sagebrush Rebellion to collapse and created a melee. Administration critics protested that previous actions had been a smoke screen to allow the unmitigated theft of the lands. Among rebel ranchers, the fear arose that corporations or foreign interests might purchase rangelands out from under them.[24]

Adding to the administrative chaos, Watt himself was under attack from the Real Property Initiative group. Hanke wrote a letter to the *New York Times*, asserting that the interior secretary was irrelevant because he was not acting "in an authentically conservative way." He should, Hanke declared, raise park fees by 2,000 percent and privatize some of the system's seventy-three million acres.[25]

Under intensifying pressure, in July 1983 Watt made an infamous gaffe that cost him his job. Speaking to the U.S. Chamber of Commerce, he described the panel assembled to review problems with his coal-leasing program, saying, "We have every kind of mix you can have. I have a black, I have a woman, two Jews and a cripple." The administration tried to ignore the national outrage when the statement aired on television. But as protests persisted, congressional Republicans threatened to join the

Democrats in demanding that Watt quit. On October 10, 1983, Watt sent the president his letter of resignation.[26]

Losing its two most volatile promoters did not derail the administration's prodevelopment actions. William Ruckelshaus had replaced Gorsuch and calmed the acrimony she had perpetuated. But Ruckelshaus was unable to convince the OMB to restore the EPA's budget cuts, and he left before the close of Reagan's first term. Reagan friend William P. Clark took over at Interior, and resource production on public lands continued.[27]

Other appointees worked to roll back regulatory procedures. The BLM now mandated that in environmental impact statements the planners use "no government action" instead of grazing reductions. BLM director Robert Burford later commented, "I have seen a strong bond develop between BLM and the public land users. This partnership has successfully reversed the lock-up trend of previous administrations." Environmental historian Samuel P. Hays placed a different emphasis on Burford's management, saying, "[He] took up the cause of the livestock operators."[28]

The Office of Surface Mining revised regulations and reduced enforcement. Environmental groups used lawsuits to blitz the agency, and persistent investigations by Congress led a House committee to declare the office "a shambles." Attempts to facilitate offshore drilling were met with more court challenges, but decisions in 1983 favored the oil and gas companies, allowing development in the coastal zone.[29]

Former Louisiana-Pacific Corporation attorney John Crowell, overseeing the Forest Service, set immense timber harvest levels and demanded each forest meet its share. Citizen groups, scientists, and whistle-blowing employees galvanized against the policy. A radical environmental group, "Earth First!," rebelled, using ecosabotage to stop, or at least slow, the removal of old-growth stands.

Twenty-nine environmental groups filed federal suits when the Forest Service attempted to contravene laws to allow cutting. As the 1980s progressed, Congress became more and more hostile to the agency's policies, and, at the end of the decade, a federal court issued a preliminary injunction halting 140 planned timber sales.[30]

With what was widely perceived as the exploitation of the environment, organizations dedicated to protecting federal lands made explosive gains. After adding 3,000 members between 1977 and 1980, the Sierra Club added 165,000 members, to top 346,000, by 1983. The Wilderness

Society, which had 48,000 members in 1979, grew to 333,000 by 1989.[31]

The backlash against the Reagan administration's public lands agenda ensured that there was no large conveyance of public lands. In fact, momentum from the movement against Reagan's policies pushed Congress to pass more wilderness legislation than during any other administration.

The political gap in Washington widened as an internal House Republican study, "The Specter of Environmentalism," mirrored Reagan's policies. The report suggested that the environmental movement was a rallying point for extremists, including "liberals, the counterculture, and revolutionaries." It posited that they were a growing threat to "the orderly development of the nation's resources," concluding that challenging the environmentalists would provide political opportunity. The land rebellion had become one of the driving forces in an American culture war.[32]

9

Inciting the Populace

In the 1980s, believing the environment to be under attack, left-wing radicals resorted to open rebellion, ecosabotaging projects on public lands. In the 1990s, a perceived liberalization of laws and mores, including environmental gains, created a rise in violent antigovernment rhetoric. This in turn spurred the formation of right-wing militias, increasing incidents of domestic terrorism.

On the political front, environmentalists struggled to maintain a bipartisan political approach to protecting public lands. But a number of Republican leaders had adopted the perspective that the environmental movement was extremist.

In 1994, when Republicans swept into Congress, taking control of both houses, they vowed to reverse the government's environmental policies. In opposition to the "green" environmentalists, they referred to themselves as "the browns." Opening the door to industry lobbyists, they introduced a long list of bills to dismantle environmental laws or frustrate regulatory agencies. In case after case, a small number of moderate Republicans joined Democrats to defeat the proposals; others were vetoed by President Bill Clinton.[1]

Among the era's most radical advocates for the environment was Edward Abbey. His work history included both employment with the

Forest Service and as a hand herding cattle. In numerous books, essays, and magazine articles, he used irony, bitterness, confessional deprecation, and a riotous perspective to promote the West's wildlands.

In a *Harper's Magazine* article in 1986 titled "Cowburnt," Abbey attacked the beef industry's abuse of the range. Although qualifying his statement by noting there were a few honorable exceptions, he wrote, "[The rancher] strings barbed wire all over the range; drills wells and bulldozes stock ponds; drives off elk and antelope and bighorn sheep; poisons coyotes and prairie dogs; shoots eagles, bears, and cougars on sight; supplants the native grasses with tumbleweed, snakeweed, poverty weed, cowshit, anthills, mud, dust, and flies. And then leans back and grins at the TV cameras and talks about how much he loves the American West."[2]

Abbey's 1975 novel, *The Monkey Wrench Gang*, was a virtual how-to book on attacking logging operations, dam building, or other activities perceived as despoiling wild places. *Monkeywrenching* came to be a synonym for sabotage in the name of environmentalism. Beginning in the early 1980s, small independent groups in the Rocky Mountains and the Pacific Northwest garnered nationwide publicity under the radical banner "Earth First!" In forests scheduled to be logged, they spiked trees and roadways, cut fences, disabled vehicles, and built platforms to sit in targeted old-growth trees.

Dave Foreman, who formerly worked for the Wilderness Society, cofounded Earth First! Living in New Mexico, he said he had been goaded into pursuing a new direction by personal threats and "the howling, impassioned extreme stand set forth by off-road-vehicle zealots, many ranchers, local boosters, loggers, and miners." He asserted that the traditional environmental groups had proved unable to save threatened lands or species. In 1985 he and cohorts published a compilation of articles entitled *Ecodefense: A Field Guide to Monkeywrenching*. The use of sabotage slowed projects, influenced agencies evaluating old-growth forests, produced financial setbacks for resource extractors, and created headaches for law enforcement. It also allowed antienvironmentalists to portray environmentalists as radicals, helping incite a violent backlash.[3]

Two major antienvironment and antigovernment groups joined in the fight against regulatory land policies: the Wise Use movement and the County Supremacy movement. The organizations grew as regulations tightened on rangelands and forests.

The Wise Use movement was a political vehicle. Ron Arnold, who wrote a laudatory biography of James Watt, was its outspoken founder. Arnold touted three things as primary American tenets: individual liberty, private property, and corporate capitalism. In 1987 he was instrumental in organizing a meeting that created Wise Use. The alliance included business interests, ranchers, rural workers, and middle-class taxpayers who believed private property rights were being infringed on. Because the organization's funding came largely from resource extraction interests and sought to open federal lands to them, historian Richard White described Wise Use as a "movement that seeks to mobilize rural western discontent for the benefit of corporate producers and developers."[4]

In the early 1990s at community picnics, parades, fairs, and rodeos, Wise Use posters announced "Join the Grassroots Rebellion against Over-Government" or "Fight to Keep Public Lands Open." The group's themes involved states' and private property rights, expanded rights to graze cattle on public lands, and the right to bear arms in national parks. It also lobbied for use of public lands for mining and logging. It emphasized its philosophy that Americans' most precious right was to remain free from "dangerous federal government encroachments." By the year 2000 Wise Use could claim thirty thousand members.[5]

Wise Use proponents framed environmental problems in economic terms. They asked if public concern should be for endangered species like the northern spotted owl or for the families of loggers barred from employment in the owl's habitat. Conservative media repeatedly publicized idiosyncratic scientist Dixie Lee Ray's phony assertion that because of the Endangered Species Act, millions of dollars were being spent to protect cockroaches and dung beetles. Wise Use advocates acknowledged the buildup of carbon dioxide in the atmosphere merely to promote clear-cutting old-growth forests, saying young carbon dioxide–absorbing trees would replace them.[6]

Organizer Arnold used hyperbole to demonize environmentalists and federal agencies and to excite audiences. Typical was the assertion that "the National Park Service is an empire designed to eliminate all private property in the United States." Arnold described environmentalists as Goliath, saying, "And we're David, and we intend to put the stone in their head." He also said, "We're not out to kill the f--s. We're simply trying to eliminate them. Our goal is to destroy environmentalism once and for all."[7]

Arnold's bombastic rhetoric paralleled that of the right-wing media, particularly talk radio. Hosts engaged in daily diatribes against those who promoted regulation. Right-wing radio personality Rush Limbaugh pursued an ongoing culture war against those he termed "environmentalist wackos," "commie-libs," and "feminazis." He derided animal rights groups, playing Andy Williams's song "Born Free," punctuated by gunfire and shrieking animals. Limbaugh insisted it was all in fun, and Republican politicians, using the show as a soapbox, lent it legitimacy. In 1994 Limbaugh was named an honorary Republican member of the 104th Congress.[8]

Republican politicians like House Speaker Newt Gingrich mobilized constituents by denying global-warming science and demanding that the "job-killing" EPA be abolished. President George H. W. Bush's appointees attacked clean-air regulations. Under pressure from Bush, the EPA withdrew evidence it had used to propose owl-habitat preservation, reducing protected territory from 11.6 million acres to 5.4 million. His interior secretary waived requirements of the Endangered Species Law, allowing harvests to proceed on thirteen previously prohibited timber sales.

The Republican Party's 1992 platform accused Democrats of using "junk science to foster hysteria" over the environment. "Unemployment is a form of pollution too," it declared, "contaminating whole communities." Bush lost his bid for reelection. Among the issues, the Gallup Poll showed two-to-one public disapproval of his environmental policies.[9]

In the early 1990s, right-wing politicians were making fantastic, fear-inducing claims. There were persistent accusations that environmentalists and liberal politicians sought a "New World Order," were championing the United Nations to create a one-world government, or were seeking to repeal the Second Amendment right to keep and bear arms.

In the Waco siege in 1993, more than eighty cult members and several federal agents were killed as the Bureau of Alcohol, Tobacco, and Firearms (ATF) and the Federal Bureau of Investigation tried to enforce firearms laws. Texas Republican congressman Steve Stockman claimed it was a Bill Clinton plot to ban assault weapons.

In February 1995 Helen Chenoweth, a Republican U.S. representative from Idaho, publicized the false claim that armed wildlife officials were flying black helicopters onto ranchers' property to enforce the Endangered Species Act. Before her election she had told an audience that the government was "unlawful from time to time," a statement used by a violent paramilitary group to promote recruitment. And she sought to create a law requiring federal police to get permission from local sheriffs before they could enter Idaho. Two Republican senators sent a letter to the Justice Department, raising concerns over the "growing militarization of our domestic law-enforcement agencies."[10]

Although politicians supported the Wise Use movement, as did groups like the Mountain States Legal Foundation, billing itself as the litigation arm of Wise Use, so too did militias and racist cults. Wise Use philosophies created an intellectual framework for organizations on the extreme fringe. They took the antigovernment campaign as license to foment open rebellion and violence against environmentalists and civil servants.

In the 1990s militia-connected individuals destroyed environmentalists' homes; attacked them—in the most publicized cases knifing one, raping another; bombed two Nevada public offices; and attempted to bomb other federal buildings, one in Reno, another in Seattle, and still another in Austin. They also issued innumerable death threats. Frequent target Andy Kerr, of the Oregon Natural Resources Council, somberly acknowledged that the threats "come with the territory these days." Shortly before he was to testify against clear-cutting on the Navajo reservation, Native American environmental leader Leroy Jackson, who never drank or smoked, died of an injection of methadone—with Valium and marijuana also in his blood. His friends insisted he had been murdered.[11]

Some rural counties in the West, with ranchers who felt imposed on, began promoting the County Supremacy movement. County Supremacists rejected the authority of the national government. They ignored historic rulings regarding Congress's ability to create and amend laws, contending that predominating all else in American political bands was the Tenth Amendment's reservation of rights to the states.

In 1991 Catron County, a rural New Mexico district of fewer than three thousand residents, became the first jurisdiction to claim "county rule." It asserted that it had precedence over federal environmental regulations. Included in its provisos was the notion that federal grazing permits were private property. Their manifesto outlawed the reintroduction in the county of wolves, bears, or cougars and threatened to arrest any federal official who objected to county measures. The county later passed an ordinance requiring gun ownership, stating, "Every head of a household residing in Catron County is required to maintain a firearm of their choice, together with ammunition."[12]

Some fifty-eight other counties embraced variations of the supremacy theme. One such district, Nye County, Nevada, is a sparsely populated landmass of 18,155 acres, the third largest in the United States. With the Nevada Test Site and Yucca Mountain being proposed as the nation's nuclear waste dump, 93 percent of the land is federally owned, and the government is its primary source of employment. On July 4, 1993, rancher and county commissioner Richard Carver forced a Forest Service ranger out of the way as he drove a bulldozer to open a washed-out Forest Service road that Carver claimed was county property.

Asked about the county superseding federal jurisdiction, Carver would merely tap his shirt pocket, which held a small U.S. Constitution, a customary accessory for County Supremacy advocates. Nevada's attorney general, Frankie Sue Del Papa, found Carver's legal theory to be without merit. Carver responded sarcastically by writing to all Nevada legislators that with Del Papa in charge, there was no need for a governor or legislature. In 1996, three years after the Justice Department filed suit for the road incident, the federal district court in Las Vegas ruled against the county, defining the supremacy movement as an illegal ideology.[13]

The coordinating body for the County Supremacy movement was the National Federal Lands Conference (NFLC). Ron Arnold served on the conference's advisory board. In 1994 the title of a NFLC newsletter cover story was "Why There Is a Need for the Militia in America." The article assailed government personnel, declaring that "scalawags and rascals and mischievous persons and people open to temptation and flat out liars and thieves [are] in places of power in our federal government." It claimed that officials were poised to enact a "seizure order," forcing the nation into "an absolute martial law made of repression." Referring to environmental groups, Ruth Kaiser, the NFLC's executive director, asserted that it was coordinating with other organizations "to run the other side off the map, or die in the effort." This extreme rhetoric was pushed further by associated groups that began issuing threats and engaging in violence.[14]

When the BLM canceled the Barstow, California, to Las Vegas motorcycle race across the desert because of complaints about irreparable damage to the fragile ecosystem, riders formed the "Sahara Clubbers." Their race had been sabotaged by Earth First!ers in the late 1980s. In response they threatened that if they caught Earth First!ers setting road traps, they would take care of them with baseball bats. In the early 1990s they expanded their focus: disrupting environmentalists' meetings and using "dirty tricks" to prevent protests against the cutting of redwoods.

Including government workers with protesters, the Sahara Clubbers' newsletter advised its supporters to "do whatever is necessary to fight our enemy and fight for our freedom. We don't want to know who you are, or what you're doing, but get the job done." The club labeled U.S. senators Barbara Boxer and Diane Feinstein "ultra-liberal bitches" for promoting a bill to create a national park in part of the Mojave, and Sahara Clubbers physically harassed female Sierra Club members who were trying to testify in support of the park.[15]

In 1994 militias were using Wise Use groups for recruiting. In a small village in eastern Washington, the Militia of Montana held an organizing meeting, and there was random talk of "taking people out." Jere Payton, an environmental activist, said, "There's a lot of loose talk about killing people in our community." Assertions linked Wise Use to a growing pattern of intimidation and violence. Groups began monitoring Wise Use groups' militant activities. The National Parks and Conservation Association published a column in its monthly magazine titled "Wise Use Watch." Ron Arnold's Center for Defense of Free Enterprise issued a lengthy declaration of nonviolence in 1994. While defending vigorous advocacy of Wise Use causes, the statement denounced the use of weapons, vandalism, or personal violence.[16]

The statement did not deter the most violent act of domestic terrorism in modern American history. In April 1995 the bombing of the Alfred P. Murrah Federal Building in Oklahoma City caused the death of 168 innocent men, women, and children and injury to 680 others. The perpetrators were part of the militia movement who insisted the government was waging war against its own citizens.

The day after the Oklahoma City bombing, a Catron County rancher threatened Forest Service workers who were reducing his permit due to overgrazing. "If you come out and try to move my cattle off," the rancher declared, "there will be one-hundred people out there with guns to meet you."

When the Wise Use organization was accused of inciting open rebellion "by inviting angry rural people to strike out against their 'oppressors,'"

Ron Arnold denied culpability. As for the National Federal Lands Council newsletter citing the need for militias, Arnold claimed he had resigned from that organization three years earlier. Regarding militia members, Arnold wrote them off as "mostly illiterates and bummed-out vets suffering from some kind of shock syndrome."[17]

In a speech a week after the Oklahoma City bombing, President Clinton seemed to implicate right-wing radio hosts. "We hear so many loud and angry voices in America today whose sole goal seems to be to try to keep some people as paranoid as possible and the rest of us all torn up and upset with each other," Clinton said. "They spread hate. They leave the impression, by their very words, that violence is acceptable."

Never one to give an inch, Rush Limbaugh argued that liberals intended to use the bombing "for their own gain." He added, "The insinuations being made are irresponsible and are going to have a chilling effect on legitimate discussion." Radio host G. Gordon Liddy was infamous for his involvement in Watergate and for telling listeners that because ATF agents wear bulletproof vests, one was better off shooting for the head. After the bombing he declared, "If a listener responds inappropriately it is beyond my control and not my fault."[18]

A "powerful assumption" surfaced that attacks on "big government" played a part in rousing the militia fanatics. Progressive pundit E. J. Dionne said that politicians should "examine their consciences and ask whether their approach to winning political battles may be aggravating dark passions."[19]

The *Los Angeles Times* characterized Speaker Gingrich's comments in discussing the bombing on television as "chilling in their implications and reckless in their potential effects." The paper said, "While making the obligatory condemnation of violence, Gingrich seemed to skate on the edge of trying to justify the kind of paranoid apprehensions about the federal government that can fuel the most extreme elements of the antigovernment crowd." In an interview, Gingrich "complained that the broad Republican revolution cannot be blamed for the violence of extremists who take its complaints too seriously."[20]

Some months later, domestic saboteurs derailed a train in Arizona, killing one and injuring one hundred. Two other plots to bomb public buildings in the state were foiled and the conspirators arrested.

In December, outside the IRS office in Reno, a one-hundred-pound bomb similar to that used in Oklahoma City—composed of ammonium nitrate and fuel oil—was ignited but failed to detonate. And in Nevada's Sierra foothills, a bomb exploded under the van in the driveway of a Forest Service supervisor's house, burying a living room couch in glass. The supervisor was traveling, but his wife and three children were in the room a few feet from the explosion. It was a warm night, and the windows were open. The kids were watching television and talking. The wife said, "They knew we were in there on the other side of the window." The family was unhurt, but the supervisor teared up when he was interviewed. "I don't know what I'd do if I lost one of the kids," he said. "And yet, people do things like that."[21]

Pictured on Glacier Point in Yosemite, President Theodore Roosevelt and John Muir agreed about protecting America's iconic places. Despite the fact that the Supreme Court identified Congress as the proper agent of federal action, the president took it upon himself to increase public lands from 42 million acres to 172 million. *Courtesy the Library of Congress.*

Gifford Pinchot, Roosevelt's chief of forestry, challenged timber companies, mining interests, and western ranchers, arguing the forests needed to be managed for the good of all society. *Courtesy the Library of Congress.*

Scene at the mining exchange building in Cripple Creek, Colorado, during mining unrest a year after Pinchot spoke there. Pinchot had surprised western stockmen and miners by coming to speak to them about restricting public land use. He won plaudits for showing up but not for his speech. *Courtesy Denver Public Library, Western History Collection*, X-60263.

The Grand Canyon about the time Theodore Roosevelt declared it a national monument to protect its overlooks from privatization. *Courtesy Denver Public Library, Western History Collection,* MCC-1206.

America's longest-serving secretary of the interior, Harold Ickes (*middle*), with the director of grazing, Richard H. Rutledge (*left*), passing a commemorative gavel to Edward T. Taylor, author of the Taylor Grazing Act. Ickes fought hard to pass the act in 1934, hoping to improve the range. Opposed by western ranchers and their legislators, however, federal management was largely ineffectual until the 1960s. *Courtesy the Library of Congress.*

Powerful Nevada senator Pat McCarran, whose advocacy for ranchers created recurring problems for those trying to regulate public lands. When grazing fees on Forest Service lands were thirty-one cents per animal per month, McCarran successfully kept Grazing Service fees at five cents by arguing that agency actions had not improved the range. *Courtesy the Library of Congress.*

In an eight-year battle in Congress to pass the Wilderness Act, Wilderness Society executive director Howard Zahniser never missed a hearing for the bill. After sixty-six rewrites, battling ill health, he produced the final version of the bill, although he died four months before it became law. *Courtesy Denver Public Library, Western History Collection,* Whexhibits.

Ronald Reagan at a Heritage Foundation dinner with Joseph Coors and his first wife, Holly, founding members of the foundation. When elected, Reagan set aside a task-force report on the environment by Republican administrators for a Heritage Foundation report that emphasized natural resource extraction on public lands. *Courtesy Denver Public Library, Western History Collection, RMN-027.*

President Reagan's secretary of the interior James Watt was a frequent target of political cartoonists. This 1982 *Washington Post* cartoon, by the award-winning cartoonist Herbert L. Block, features Watt driving a steam shovel through wilderness. At the time Watt was promoting a bill that purported to protect wilderness, but would have allowed the president to withdraw parts of the system, paving the way to sell public lands to development interests. *A 1982 Herblock Cartoon. Copyright The Herb Block Foundation.*

Shoshone Indians Carrie (*left*) and Mary Dann, waiting as the BLM rounded up and confiscated livestock on their ranch in February 2003. The Dann sisters refused to accept reparations for Shoshone land, pointing out that there was no treaty ceding it to the government. *Courtesy Special Collections, University of Nevada, Reno, Library.*

10

County versus Federal Government

The state of Nevada encompasses more than 70,260,000 total acres. The BLM administers some 48 million acres of that area, with other federal agencies overseeing 9 million more. A primary reason for the amount of Nevada land that remains under federal jurisdiction is that the Great Basin, which encompasses all of Nevada, half of Utah, and parts of the adjoining states, has a fundamental shortage of water. No great rivers run into Nevada, and major storms, which generally form in the Pacific Ocean, lose much of their precipitation in the Sierra Nevada on the state's western boundary. In the rain shadow, depending on elevation, the Great Basin averages only between three and twelve inches of precipitation a year.

The first settlers and those in the early twentieth century, claiming acreage under the Stock-Raising Homestead Act, took what watered lands there were. Nevada state senator Dean Rhoads pointed out, "They were very wise, those old-timers. They'd go up a creek, and they'd take [the areas] like a snake, and that would be their 640 acres." The sparsely forested mountains and desert lowlands were ignored. In recent history, use of the remaining federal lands has been particularly contentious.[1]

A July 4, 2000, incident was the climax of twenty years of calls for open rebellion in Elko County. Those who gathered to reopen a washed-out, unpaved Forest Service road, near the tiny town of Jarbidge, called themselves the "Shovel Brigade." The brigade leaders included a county commissioner, a fourth-generation rancher, and the head of the county Republican Party. Also taking part was a Shoshone Indian trying to publicize that the 1863 Ruby Valley Treaty proved the tribe had never ceded their land to the United States. Supporters of the Shovel Brigade, estimated to number between 300 and 750, came from several western states and from as far away as Rhode Island and Georgia.

Elko County's history of clashes over federal jurisdiction included the designation of the Jarbidge Wilderness in the early 1970s. At the time county leaders argued that the Forest Service was more interested in protecting mountains than "preserving the freedoms of the American people." Another fight erupted in 1995 when a rancher claimed rights to a spring on federal land. The county defended him. Community members built a fence around the site and posted a sign saying that the land and water inside belonged to the people of Nevada. Before the rancher finally agreed to obtain a permit to access the water, the county had spent $450,000 on legal fees.

The Jarbidge road, a couple of miles long, ended at a campground on the edge of the sixty-five-thousand-acre wilderness. It was located along a creek that flooded in 1995. The Forest Service decided not to rebuild the road, as it carried silt into the creek, harming endangered bull trout.[2]

For those resentful of the federal bureaucracy, the truncated road became a larger cause than the spring on public land. The county commissioners decided to repair the road without obtaining the appropriate permits. To begin they filled in nine hundred feet of wetlands, changing the course of the creek. The U.S. Army Corps of Engineers and the Nevada Department of Environmental Protection got an injunction against the county for violating the Clean Water Act. Two local politicians and the county attorney were determined to continue the work. A restraining order sidelined them, but they continued to rally others to the cause.[3]

In February 2000, at a Republican gathering in Las Vegas, at the opposite end of the state, Nevada's U.S. senator John Ensign presented shovels to two of the leaders. At the meeting Ensign, who was later forced to resign from the Senate for ethics violations, lectured those he dubbed "easterners" about the states' rights "enshrined in the Tenth Amendment." He also attacked the government, saying, "I will tell you flat-out today that most of what the United States government does is unconstitutional. If we don't stand up out here in Elko, across Nevada and across America, they are going to continue to get more and more power and give less and less liberty."[4]

The antigovernment campaign continually spurred animosity toward those charged with upholding regulations. Forest Service personnel and their families were the frequent targets of insults and intimidation in Elko. Children were threatened and ostracized at school, and wives were refused service at stores.[5]

Bill Kohlmoos, the president of the Nevada Miners and Prospectors Association, dismissed the effects of the bullying, saying, "A forest ranger was insulted in Elko, or a forest ranger couldn't get a motel room at a certain motel, or one of the gas stations wouldn't sell gas to a federal employee. Well, so what? Those are small things. . . . Heck, two people go to the post office at the same time and they have a meeting in the doorway. One of them might say, 'Get out of the way,' or something. I mean, so what? It doesn't mean anything."[6]

To the federal employees, it meant a great deal. In several national forests in the West, supervisors had had to change policies to better protect rangers. Gloria Flora, a twenty-three-year Forest Service veteran and the forest supervisor of Nevada's Humboldt-Toiyabe, the largest forest in the lower forty-eight states, resigned her position in protest over the threats to her employees. "The attitude towards federal employees and federal laws in Nevada is pitiful," she said, blaming politicians at all levels in the state for "fed bashing." She bemoaned that people in rural areas who respected the law "are compared to collaborators with the Vichy government in Nazi-controlled France." Explaining defensive changes instituted before her resignation, she commented, "The fact is that you're vulnerable. Yes, many officers got rid of the green vehicles. Many of us stopped wearing our uniforms. We did not permit people to go out alone in the field; always travel in twos. Hunting season we were particularly cautious. We increased our radio coverage, improved our radio communication."[7]

Under those circumstances, at the July 4 act of civil disobedience in Jarbidge, an all-terrain vehicle (ATV) pulling a trailer with a tombstone labeled *Forest Service* did not seem funny. Demonstrators drove or rode in rented Idaho school buses thirty miles over a washboard gravel road to the creek outside Jarbidge. With a banner proclaiming, "I love my country but fear my government," the protesters sang "The Star-Spangled Banner," and some two hundred people grabbed heavy ropes and dragged aside a four-ton boulder used by the Forest Service to block the road.

Two dozen members of the environmental group Great Old Broads for Wilderness, there to protest the reopening of the road, kept their presence low-key. They were there to assert that the wildlands, wild water, and wildlife—in this case, the bull trout—needed representation. Their leader, Susan Tixier, commented, "The county commissioners in Elko County have demonstrated—and other counties across the West—they are more interested in the constituents who have elected them, who are their friends and relatives, than the federal public lands." Rumors had spread that as many as ten thousand demonstrators could be at Jarbidge, and the Great Old Broads contingent had been warned that it might be dangerous. Even with the crowd in the hundreds, along with the mock tombstone there were other threatening displays: a sign declaring "Tree Huggers: the other red meat," a heavily armed man riding among the crowd on a horse, and at the top of a hill a huge picture of a bull trout, "its head all shot up." Referring to their decision to remain in the background, Tixier said that originally the demonstration seemed "a little terrifying," but that it turned out to be more like a county fair.[8]

A probable reason the crowd remained in good spirits was the absence of uniformed authorities. The U.S. Attorney's Office had threatened to arrest protesters, but they stayed away at the request of the county sheriff, who assumed their presence would "set off fireworks." Also absent was Matt Holford, the Nevada director of Trout Unlimited, who complained, "With the next rain, all that [road] material is going to be back in the river." Holford, receiving a decidedly cold reception in Jarbidge the preceding day, had withdrawn.

After the boulder was moved, Jarbidge's ninety-year-old Helen Wilson took the first ride down the road, in a pickup truck. "To heck with the fish. They're slimy and soft and they don't eat well," she said. "This is all politics." The Great Old Broads' Tixier conceded that her group had been "out-grandmothered."[9]

Because the work took place outside of the approval process, the government sued the local officials, and the animosity between the county and federal agencies continued. In 2003 the Shovel Brigade, still presided over by a county commissioner, placed an advertisement in the *Elko Daily*

Free Press and on Elko radio stations, identifying BLM and Forest Service employees as "armed and dangerous." The ad blamed armed officials for creating several nearly explosive incidents by attempting to enforce laws or confront armed citizens. It repeatedly called the federal officers "gunmen" and concluded, "The Brigade thinks it is time for a congressional review of the policy which allows gunmen to roam around the counties as representatives of the bureaucracies."[10]

As the reconstituted Jarbidge road washed out in subsequent winters, negotiators for the two sides agreed to settle the lawsuit and allow the Forest Service to build a more environmentally friendly road. But the Elko County commissioners rejected their negotiators' agreement, demanding the federal government deed the road over to them. An appeals court allowed the Wilderness Society and the Great Old Broads for Wilderness to try to block any subsequent deal, as they sought to have the road scaled back to a trail open only to nonmotorized use.

Not everyone in Elko has supported their commissioners. A retiring county clerk asserted that the county was broke from fighting the federal government's management decisions. "It's the same radical people with the same radical words," she said, "and they want us to foot the bill." In 2016, after sixteen years and the county having spent $250,000 on Jarbidge, the case was still being litigated.[11]

11

Extinguished Rights

The Forest Service problems in Nevada were mirrored by law-enforcement troubles the BLM faced. In western land battles, individuals, like those in the Shovel Brigade, seek to replace federal managers' judgment with their own. Ranchers worry about losing their way of life as they attempt to balance their needs against national interests. The individuals risk their property, and at times their liberty, in attempting to maintain their relationship with the land. Native Americans, with their original claims, present a particularly difficult dilemma for land agencies.

In spring 1992 Clifford Dann, a Shoshone Indian in Crescent Valley, Nevada, pulled his truck across a road to block a BLM convoy of vehicles full of his family's livestock. The confrontation took place after a six-day standoff. As county sheriffs and federal agents stopped their vehicles, Dann, standing in the pickup bed, doused his arms with gasoline, produced a lighter, and threatened to immolate himself. "By taking away our livelihood and our lands you are taking away our lives," he declared. Carrie Dann, Clifford's sister, in her fifties, accompanied by several Indian and non-Indian supporters, began arguing with the officials. When told none of the covered trucks contained Dann cattle, Clifford Dann got out of his pickup bed to see for himself. He still carried the gas container and lighter when an agent grabbed him from behind. Other officers rushed him, spraying white foam from fire extinguishers and wrestling him to the ground.

Carrie Dann was held back by officers as her brother was taken away, accused of assaulting a federal officer with gasoline. Carrie had prevented a previous attempt at confiscation by physically blocking animals from being loaded in trucks. She had demanded to see documentation of a Shoshone land transfer to the U.S. government. This time asking for evidence had no effect.

The only agreement between the Western Bands of the Shoshone Nation and the United States is the 1863 Treaty of Ruby Valley. It was a peace treaty with no mention of land rights. A year after Dann's arrest, the Nevada District Court rejected the argument that the treaty proved the Shoshones had never signed away title to their Great Basin lands.

Clifford Dann refused to defend himself further than that, saying it would be accepting the court's jurisdiction over Shoshones on Shoshone land. A jury found him guilty of assault, and Judge Howard McKibben handed down a sentence that required the Indian elder to spend nine months in Nevada's maximum security prison. McKibben handed down the stiff sentence "to send a message to journalists, activists, and the Western Shoshone."[1]

The incident, together with a similar impounding of another Shoshone's livestock, resulted in 385 horses being taken and sold in 1992. It was not the first or last incident involving the Shoshone Indians or the Dann family. There were similar livestock confiscations: in 2003 the Danns lost 232 head of cattle. Other instances of Shoshones being jailed included arrests for not complying with state hunting laws. In the latter cases, tribal members contended they were engaged in subsistence hunting on ancestral lands.[2]

The Danns' troubles began in 1974. Carrie's older sister, Mary, was out checking on their cows. When she returned home, a BLM official was waiting. He told her that neighbors had complained about Dann cattle straying onto land for which they had no grazing rights. "Do you know you're trespassing?" he asked. She answered that the only time she considered herself to be trespassing was when she went over into Paiute land. She used a map to point out the immense traditional Shoshone territory, stretching from the Great Salt Lake to California's Death Valley. She demanded the official produce documentation that changed the land's status. Citing the fact that there was no such document, the sisters refused to pay grazing fees.³

For the ensuing decades, Carrie and Mary Dann, while running their ranch, engaged in disputes to establish that their grazing land is Shoshone. Their claim was buoyed by the fact that the 1863 treaty recognized the indigenous land base, merely granting the U.S. access for limited, specified purposes.⁴

The treaty had been eagerly sought by the government. The violence between the Shoshones and Americans was a stumbling block to Nevada mining and the construction of the transcontinental railroad. The mines' immense deposits of silver and the ability to transport the riches were key components for the Union in the Civil War. Access to the ore was a primary factor in convincing Britain and France that there would be no disunion, thus keeping the Europeans from assisting the South.⁵

Western Shoshone land-claim problems came to the fore in 1951 when some tribal members, encouraged by the Bureau of Indian Affairs, filed a claim with the Indian Claims Commission. Congress formed the ICC in 1946 to provide a means for tribes to be compensated for land and resources taken from them. Because no land would be returned, tribes were allowed to file suits in attempts to gain monetary compensation. There were some five thousand Shoshones living on reservations, in allotment areas and colonies, on ranches, or in towns across northern and central Nevada. Their views on land claims were as disparate as their residences. The ICC claims process drove them into two camps, those favoring land retention and those seeking financial remuneration.

The two sides battled into the 1960s, with the tribal attorneys representing those seeking monetary compensation. One of the tribe members seeking payment said, "The money is earned, and could be used for college education and home improvements and hospital bills." The Shoshone traditionalists asserted that no one could buy land, that it, and the creatures it harbors, is sacred. Traditional leader Raymond Yowell explained, "The land cannot be sold. That is our religious belief. The Creator placed us here as caretakers of the land and he didn't say that at some point you can sell this for money." Carrie Dann said, "I wouldn't take a million dollars per acre. If I did, I would be selling my pride, my honor, my dignity, my birthright, everything that says I'm a Western Shoshone.'"⁶

In 1962 the ICC had ruled that by "gradual encroachment," the Shoshone lands had been "extinguished." Lacking any date that the sparsely occupied land had been taken, the ICC stipulated July 1, 1872, as the "date of valuation." At that time, there were miners, a few towns, and a thousand small farms scattered throughout Nevada. As with the entire paternalistic ICC approach, its scrutiny of the area excluded input from Shoshones. Appraisers methodically tallied the values of town sites, agricultural lands, and grazing lands in 1872 as well as mineral profits up to the date of the taking. Protestations by traditional members of the tribe caused lengthy delays. Finally, in December 1979, an award of $26 million was made for dispensation to the Shoshones, if ever the disputes were settled. The valuation came to approximately 15 cents an acre.[7]

With tribe members still battling among themselves, the attorneys were paid their fees, $2.6 million. Regarding the principal attorney, leader Yowell commented, "For over twenty years, Mr. Barker has been the cause of much fighting between ourselves and has done nothing to preserve or enforce our rights except try to give up our rights for money so he can get his ten percent."

In May 1983 the Ninth Circuit Court of Appeals issued a ruling that disagreed with the ICC determination. The Ninth Circuit stated that, as a matter of law, Shoshone land rights had not been extinguished. The government appealed that decision to the Supreme Court, which, in February 1985, found that payment for "taken" land had been made when the funds were deposited in the Treasury. It ignored the question of how the ICC acquired legal authority to extinguish the Native peoples' land rights. And it disregarded the fact that gradual encroachment is not usually sufficient cause to do so.[8]

The Shoshones were also stymied in pursuit of the opposite tack. They argued that if a taking of their land had occurred, the Fifth Amendment required that interest be included in the compensation. The year of the taking was 1872, and the amount was $26 million, creating an amount of accrued interest totaling $14 billion. The claim was denied. The Supreme Court ruled that when the United States took the land, the tribe did not hold fee title, or absolute legal possession. The Court declared that the Ruby Valley Treaty did not acknowledge landownership. Instead of recognizing "any exclusive use or occupancy right or title of the Indians," the treaty was "a treaty of peace and amity with stipulated annuities for the purposes of accomplishing those objectives."[9]

Carrie Dann's response to the courts' decisions was: "We never left our land; the United States can claim anything it wants, but the reality is that we're still here." She and her sister's subsequent actions left the BLM in an impossible position. As long as the Danns complied with the agency's regulations, they were eligible to graze their livestock on what the courts ruled was federal land. Their father had done so until his death in the 1960s, after homesteading 160 acres in the early 1920s. Carrie Dann said he had followed the rules, believing he would be run off the land if he did not. But she and her sister had been taught by their grandmother that "all these lands are Shoshone lands, and that the government could not make rules for us."[10]

How to pursue their claim was the problem. The sisters sent a letter to Manuel Lujan Jr., George H. W. Bush's secretary of the interior. They asked rhetorically if the government understood the relationship that traditional Shoshones have with the land. The letter was passed along to the Bureau of Indian Affairs, where an official dutifully, and unhelpfully, responded, "My staff informs me that many Western Shoshones have a strong attachment to aboriginal lands and consider them sacred."[11]

The sisters were grazing cattle and horses mixed with large numbers of wild horses. In 1991, before beginning to confiscate Dann livestock, Billy Templeton, the BLM director for Nevada, said, "I don't want a confrontation. But we have been very patient. . . . The court case is closed and yet the Danns are continuing with excessive, unauthorized use of the land."[12]

In 1992 at the Reno Federal Courthouse, Judge Bruce Thompson of the Ninth Circuit Court of Appeals sat in judgment of the Dann sisters. He had been involved in their case since the 1970s. In 1986 he had ruled that the Danns possessed individual rights to graze their stock on the Shoshone aboriginal territory. The decision did not satisfy the sisters. It meant they could use the land, but they did not own title to it. In 1992, as previously, the women were polite but firm, telling the judge he had no jurisdiction over them. Calling the history of the government's relationship with aboriginal people "shameful," the judge said, "Anyone who has read this history knows it's a history of broken promises and broken dreams. [But] all the avenues which were open from time to time for miscarriages of justice have been closed. There is nothing more I can do."[13]

Continuing to pursue traditional Shoshone rights, but failing in efforts to get the federal courts to again hear their case, in 1993 the Danns lodged a petition with the Inter-American Commission on Human Rights. Finally, in December 2002, the commission concluded that the government had failed to protect the Danns' right to equality before the law, the right to judicial protection and due process, and the right to property. It recommended that the United States provide the Danns with an "effective remedy, which includes adopting the legislative or other [necessary] measures."[14]

By that time the Danns' herds along with wild horses amounted to some 1,500 animals on lands that the BLM contended should feed 180. In 2002 members of a citizens advisory group visited the ranch and expressed disgust. BLM field manager Helen Hankins said, "It was just dirt." She said deadlines for a settlement had been missed and that meetings with the Danns went on for four or five hours with nothing ever resolved.

When asked if they had been good stewards over recent years, Carrie Dann responded, "I can't say that." She qualified her statement by saying, "They've forced us into this position." The *they* may have been referring to mining interests as well as the government. Through U.S. mining legislation and property exchanges, the lands around the Danns' ranch and grazing area were being impacted. Large-scale, intrusive mining had begun in some areas, gold prospecting in others. The industrial activities included both pumping scarce water out of the Humboldt River drainage and contaminating the groundwater.

The Danns' problems escalated with the results of a Shoshone straw poll in 2002. Organized by the chair of four Shoshone bands, the vote supported taking money, 1,647 in favor to 156 against. Most "no" voters simply did not take part.[15]

With the tribe still split on indemnification, in 2007 the Bureau of Indian Affairs began accepting applications from Shoshones to receive payment for their taken lands. Because of interest accrued over more than thirty years, the total government payout was some $188 million, one of the largest land-settlement claims in ICC history. Those who filed would receive more than $35,000 each. Distribution began in 2011. Mary Dann had died in 2002. Carrie insisted that the land was life for future generations and that those taking the money were "ripping them off."[16]

Through the years the sisters won several human rights awards, including a Petra Fellow Award in 2003. But as for the Danns' legal case, international advocacy reinforcing the family's physical resistance has done nothing to induce the judiciary to reexamine previous decisions.[17]

12

Pursuing Ideology in the Courts

Wayne Hage, who died in 2006, was one of the intellectual leaders of westerners advocating for property rights. Most ranchers accepted the relationship between their actual property, which included ranch land and water rights, and federal grazing lands. Their complaints dealt with the volatility of regulations since the implementation of multiple-use or fee increases. Hage challenged the basis and nature of ranchers' dependence on the public range. His fight with the government regarded the Fifth Amendment's delineation of takings: No person shall be deprived of property without due process, "nor shall private property be taken for public use, without just compensation." He argued that the Forest Service's mandate to protect resources was, in fact, a strategy to hide taking a rancher's property. The battle lasted twenty-five years, continuing after his death.

In 1978 Hage bought the Pine Creek Ranch below eleven-thousand-foot Table Mountain, near Tonopah, Nevada, for about two million dollars. The 7,000-acre ranch included a number of water rights, and he applied for and was granted grazing rights on 752,000 acres of Forest Service and BLM lands. Hage had a master's degree in biological science. He had served on boards and committees, including the Agricultural Land Use

Committee of the California Chamber of Commerce. He also sat on the executive board of the Mountain States Legal Foundation, which was involved in numerous lawsuits challenging federal regulations.[1]

Shortly after taking possession of the ranch, Hage began challenging Forest Service rulings. In 1978 the agency, noting unauthorized grazing, requested that Hage move his cattle. Various grazing violations would be repeated throughout the 1980s. In 1983 alone the rancher received approximately forty letters and seventy visits over charges of public land abuse.

Hage had his own objections, in particular that the Forest Service had allowed the Nevada Department of Wildlife to introduce nonindigenous elk into the range. He complained that the elk competed for forage and water and elk hunters scattered his cattle. In central Nevada the Forest Service would erect small enclosures around the heads of springs to protect them from being trampled and ruined. By keeping the spring from caving in, the water was allowed to run out, and the cattle could drink without stopping the flow. Hage protested that Forest Service fences kept his cattle from the water. He also contended that the agency allowed willows to grow along irrigation ditches, which usurped water from his sources. That action, he asserted, constituted a taking of his water rights. The two sides were unsuccessful in repeated attempts to resolve the issues.[2]

In 1989 Hage wrote a book, *Storm over Rangelands: Private Rights in Federal Lands*. The book was published by Wise Use movement founder Ron Arnold's Free Enterprise Press. The thesis of the book was that ranchers had valid property rights in federal lands that had not been acknowledged.

Hage contended that environmentalist leaders controlled government agencies in "a well-planned, superbly orchestrated attempt to destroy the property rights of western interests." He believed "scientific range management" was a political tool used by the Forest Service from its earliest days to push stockmen into a corner. He contended that early in the twentieth century, by creating laws, northeastern financial, industrial, and political entities had restricted the West, keeping it in "quasi-colonial status." According to Hage, the status was perpetuated when the northeasterners put Franklin Roosevelt in office.[3]

In 1990, because of overgrazing in Meadow Canyon, a section of Hage's federal land allotment, the Forest Service directed Hage to graze his cattle elsewhere. Hage filed an administrative appeal to stay the requirement, but the stay was denied. He tried unsuccessfully to move the cattle, later testifying that it was impossible to keep the cattle from the canyon because of a twenty-five-mile, mostly unfenced, boundary.

At about that time Hage, hired a woodcutter, and they used heavy earthmoving equipment to remove trees in a fifty-foot swath along each side of a ditch on federal lands. They sold the wood as commercial firewood. The Forest Service had previously notified him that a special-use permit was required before removing natural features. He was subsequently found guilty of removing federal property. But the conviction was overturned on appeal because the value of the property damaged had not been established at trial. The Hage–Forest Service disputes came to a head when, after sending at least two notices of intent to impound cattle found in Meadow Canyon, the Forest Service confiscated and sold the animals.[4]

Hage described the agency's actions in rounding up the cattle as egregious. He said they acted because he had not responded the way they wanted after they sent him the notices. On a morning Hage sent his crew up to Meadow Canyon to round up cattle because he was selling off the herd, armed range-management personnel confronted them. The officials threatened to arrest anyone who attempted to gather the cattle.

That same morning a Forest Service official from Washington, DC; another from Ogden, Utah; and a representative of the Nevada Cattlemen's Association had come to mediate the conflict. Hage drove them around, showing them areas where "we had evidence of harassment, vandalism, et cetera, et cetera." He refused to enter into any agreement. Saying it was too late for negotiations, Hage declared, "You come here with your regulations, your harassment, with your vandalism, and you have destroyed this ranch as an economic unit, and so this is going to the United States Court of Federal Claims."

In Meadow Canyon the federal employees had prevented Hage's crew from unloading their horses. Hage said that one of them, recognizing Hage's son, told the fifteen-year-old if he said anything, they would handcuff him to a pickup truck. Hage commented, "I think he began to understand that government is not your friend. Of course, he had watched the Forest Service for years. . . . He was pretty well aware of the fact that the Forest Service is a very, very corrupt organization with a criminal mentality."[5]

In 1991 Hage filed a lawsuit against the United States, alleging it owed compensation for a Fifth Amendment taking of property and breach of contract. The battles over his contention were fought in multiple hearings and resulted in numerous decisions before what appears to be a final judgment was issued in January 2016.

Hage's book, published two years before he filed suit, lays out the arguments that his case would pursue. There were two questions to be resolved: the legal status of ranchers' preexisting property rights and to what degree they are protected under the Fifth Amendment.[6]

Hage realized that in the case of *Hunter v. United States*, in 1967, the court had determined that Roy Hunter's water rights did not establish property rights. Hunter had proprietary rights to water on national monument land. But his claim to a right to graze cattle on the monument land was found to be illegitimate. The court said the government had the right to cancel the approval of land use at any time. Hunter could remove the water to a secondary site, but he was not entitled to other easements.

Hage felt that in the period when the *Hunter* decision was handed down, Fifth Amendment guarantees were being loosely interpreted. He believed that the decision erred in upholding the taking of private property. In his book Hage posited that "the opinion in *Hunter* may not withstand a well-constructed effort to overturn it."[7]

Hage's basic contention, in opposition to the *Hunter* decision, was that a grazing allotment is personal, private property. He proposed that "the rancher is not renting this ground. The rancher owns this ground." His original argument was that because stock water rights were recognized by the courts as real property, the preexisting rights to graze had equal substance as a property right. He asked, "If the pre-existing water is valid, how can the grazing right upon which it depends for its validity not also be a valid pre-existing right?"[8]

The Court of Federal Claims, where Hage filed his initial suit, hears claims against the U.S. Treasury for amounts of more than $10,000. Before 1990 property claims were rarely brought to the CFC. But that year plaintiffs won awards in two cases involving property takings, and by the end of the decade hundreds of cases were pending before the CFC. The court's chief justice was Reagan appointee Loren A. Smith, a self-described libertarian and conservative. Smith had served as chief counsel for the Reagan presidential campaigns in 1976 and 1980.[9]

Over the years Hage's CFC case would involve numerous hearings and seven opinions by Judge Smith. In 2002 the judge found that the issuance of grazing permits does not convey legal rights to the land, but he confirmed that Hage had vested water rights. The water rights could be traced back to the July 26, 1866, act RS 2477, which permitted irrigation-ditch rights-of-way across public lands.

There are two kinds of unlawful government takings: physical appropriations and regulatory actions that affect or limit property use. Judge Smith ruled that the government had effected both kinds of taking. The construction of the fences around springs created a physical taking. Requiring Hage to seek a special-use permit before removing vegetation from ditches effected a regulatory taking. Further, the agency's actions, including threatening to prosecute Hage, had harassed and intimidated him, keeping him from accessing his ditch rights-of-way. In June 2008 the judge issued his final decision. He ruled that the government owed the Hage estate—both Hage and his wife had died—more than $4 million. With interest accrued from the time of the taking, the resultant damages amounted to $14,243,542. The government appealed.

In July 2012 the Ninth Circuit Court of Appeals issued its decision reversing the major parts of the CFC finding, including damages. The finding of a physical taking was determined to be flawed. There was no evidence that the fences prevented water from reaching Hage's grazing lands. Nor had Hage alleged that there was insufficient water for his cattle on their allotments.

Regarding the regulatory taking, the government agencies have the authority to manage the federal land. The Forest Service was within its rights to require a permit to operate heavy equipment on national forestland. The court of appeals found that the grazing disputes had not stopped the Forest Service from granting Hage special-use permits. The arguments had been ongoing since 1978, but the Forest Service granted every special-use permit for which Hage applied. There was no evidence that the disputes would lead the Forest Service to deny permits after 1986, when Hage unilaterally decided to stop applying.[10]

The Hage legal team appealed to the Supreme Court. They were supported by briefs submitted by the Mountain States Legal Foundation and the Pacific Legal Foundation. But on June 17, 2013, the Supreme Court rejected the arguments, letting the ruling of the court of appeals stand.[11]

At the same time the Hage case moved through the appeals process, the estate's representatives had filed suit in a Nevada federal district court. As in the CFC, the district court had a judge who looked favorably on property rights claims. Appointed by George W. Bush, Judge Robert Clive Jones was on record as having a bias against government agencies. Jones had told Hage, "You have a court that's very receptive and sympathetic to your claim." At a pretrial hearing, the judge stated, "In my opinion, not only in this case but in many cases, the government has been all too ready to—in the name of revoking or suspending or limiting grazing licenses, the government has been all too ready in the history of Nevada to impair otherwise suspected and substantiated rights of landowners."

When the trial opened, Jones declared, "The Bureau of Land Management, you come in with the standard arrogant, arbitrary, capricious attitude that I recognize in many of these cases." He then stated: "It's my experience that the Forest Service and the BLM is [sic] very arbitrary and capricious." And he continued, directing a comment to the officials: "Your insistence upon a trespass violation, unwillful—your arbitrary determination of unwillfulness [willfulness] is undoubtedly going to fail in this court."[12]

The trial lasted twenty-one days. Judge Jones's opinion ran to 104 pages. In it he asserted, "The government's actions over the past two decades shocks [sic] the conscience of the court." He then detailed what he termed the government's vindictive actions against the Hage family. Jones concurred with Hage that although the rancher had grazed cattle on federal lands without a permit, water rights provided a defense to claims of trespass. Contravening the *Hunter* precedent, Jones proposed that Hage was entitled to a half mile of forage around and adjacent to his water rights. Jones also held a BLM official and a Forest Service official in contempt of court for interfering with Hage's water rights and inviting other parties to apply for those rights.

In mid-January 2016 the Ninth Circuit Court of Appeals vacated in part and reversed in part the Jones judgment on its merits. They also reversed the findings of contempt, saying that "the district court did not have jurisdiction over the defendant's water rights" and that the officials' actions were within the scope of their statutory duties. The appeals court charged that Judge Jones's actions "grossly abused the power of contempt." It went on to cite other cases where Jones had exhibited antagonism and bias toward government agencies and officials and noted that the judge's "arrogance and assumption of power" may qualify as conduct "properly addressed by the Judicial Council." And it took the extreme action of directing the chief judge of Nevada's district courts to remand any further action to a different federal judge.[13]

After the reversal by the court of appeals, Wayne Hage Jr. said he does not know what the future holds. He commented that an appeal to the Supreme Court would be difficult. "It was a big disappointment, not just for my family but for the entire [ranching] industry," he said. Although he is not involved with a militia himself, he went on, "It looks to me like the 9th Circuit just swelled the ranks of the militias." This comment may have been prompted by an armed standoff against federal agents going on at that time in Oregon. It involved protesters and militia members led by the sons of the County Supremacist Cliven Bundy.[14]

13

Pursuing Ideology with Guns

Cliven Bundy's 2014 insurgency, discussed in this book's introduction, was a highly publicized, dangerous confrontation that remains unresolved at the time of this publication. Bundy is a County Supremacist, declaring that he would follow county regulations but not federal laws. It is ironic that Clark County, where Bundy's ranch is located, was a major factor in bringing about his difficulties.

Clark County's political power is centered in Las Vegas. In 1989 the city, bounded by federal land, was growing as fast as any in the United States. The Mojave Desert tortoise, living in southern Nevada, had recently been listed as an endangered species. Any further expansion of the city would need to take place on BLM Mojave Desert lands. The county acquired permits to expand into federal lands by agreeing to help fund preservation of habitat for the desert tortoise in outlying areas. Some of those areas were grazing lands used by ranchers, whom, if they were willing, Clark County officials agreed to help buy out.

Some of the ranchers agreed to sell their rights. Bundy, whose ranch is seventy-four miles northeast of Las Vegas, called regulating for endangered species "wacko environmental stuff." He did not give up his own land or his grazing rights on federal land. Instead, he demanded that the

Clark County sheriff protect him and his cattle from enforcement of federal edicts, including fee collection. "Don't we pay [the sheriff] to protect our life, liberty and property?" Bundy asked.[1]

At a news conference the day after the armed standoff, Bundy called on county sheriffs across the country to "take away the guns from the United States bureaucrats." Bundy declared that if sheriffs did not take responsibility, it would fall to citizen militias.[2]

A short time after the debacle, public figures who had publicly sided with Bundy abandoned him as he employed his newfound celebrity to espouse his viewpoints and philosophies. One of his ideas was that African Americans might "be better off as slaves, picking cotton and having a family life."[3]

The Southern Poverty Law Center's Mark Potok characterized Bundy's actions and those of the militias he assembled as equivalent to an armed rebellion against the federal government. Potok worried they would influence others to do the same kind of thing. His concern was well founded, as one incident followed another in succeeding months.

In May, less than a month after the confrontation, one of Bundy's sons traveled to Utah to join San Juan County commissioner Phil Lyman's protest against the BLM. Encouraged by a flag-waving crowd, the forty-one-year-old Ryan Bundy, Constitution in shirt pocket, led a loud parade of fifty ATVs and Jeeps through a canyon on a dirt trail. The trail had been constructed illegally seven years earlier, winding over ground littered with potshards below ancient cliffside dwellings. The BLM had closed the area to protect the archaeological record of Ancestral Puebloans.

When a local sheriff deputy on the scene of the illegal ride was asked if he had attempted to stop the demonstration, he laughed and said that was the job of the BLM, but none of their officials could be found. The county sheriff said BLM officers were present, but, in order to avoid a clash, they did not wear uniforms.[4]

In June 2014 two militia members who had been turned away after a short stay at the Bundy ranch assassinated two Las Vegas police officers. The policemen had been eating lunch at a pizzeria. The killers draped a "Don't Tread on Me" flag over their victims, killed another individual, and when surrounded by police killed themselves. The same month, in California, a man who used Facebook to praise the Bundy revolt shot a BLM ranger who had asked him to move his illegal campsite. On July 1 a group of miners illegally dredged a BLM-managed section of Idaho's Salmon River with the express purpose of driving the EPA from the state.[5]

In April 2015 more than one hundred militia members flocked to Grants Pass, Oregon. They were supporting miners who refused to file a BLM plan of operation for previously unreported mining activities. As the protesters, many openly carrying firearms, gathered outside their office, the local BLM officials closed their doors and sent their employees home. As in the other cases, the intimidating behavior went unprosecuted.[6]

The following month the nonprosecution of cases came to an end. San Juan commissioner Lyman and a fellow board member were put on trial for organizing the ride through the Pueblo archaeology site in Utah. A jury found the commissioners guilty of conspiracy to ride and riding motorized vehicles on closed public lands, federal misdemeanors. Each charge carried possible penalties of a year in jail and $100,000 fines. In December 2015 the two men were sentenced to ten days in jail, thirty-six months of probation, and $1,000 fines. They also were required to split payment of $96,000 to repair the damage the ride caused.[7]

Another incident would also result in federal charges after weeks of making national news. On January 2, 2016, what would be a forty-one-day armed militia action, costing taxpayers $1.2 million, began at a Harney County, Oregon, wildlife refuge. It led to prosecutions of some two dozen militants and the death of one.

Cliven Bundy's sons Ammon and Ryan led the insurgents. Their revolt began as a protest over the imprisonment of an Oregon father and son, ranchers and community leaders, convicted of lighting fires that damaged federal land. Alleged to have set eight fires over twenty years, they were convicted of setting two, one of which was purportedly done to cover up evidence of poaching deer. The two had served relatively short sentences, but the government appealed, and they were ordered to finish mandatory five-year terms.

As in the previous showdowns, the extremists, most of whom were heavily armed, came from various states. A couple were ranchers, but like Ammon Bundy, who lives in a small Idaho town, and his brother Ryan, from Cedar City, Utah, most were not.

Ryan Payne, who had been an organizer of the 2014 Bundy ranch standoff, insisted the Oregon ranchers were being "brutally oppressed." Coming from his home in Montana, he said the militia had to give support because of federal intimidation of the people of Harney County. He declared, "It's like asking a sheep to all the sudden defend itself from the wolves." The Hammonds rejected the militiamen's assistance, turning themselves in to serve the rest of their sentences. After marching in the town of Harney, the protesters moved to a federal wildlife refuge building some miles away, and the tenor of their demonstration changed.[8]

Dubbing themselves Citizens for Constitutional Freedom, they blockaded roads and stood guard carrying military-style weapons. Lookouts with AR-15 rifles took positions atop an old bird-watching tower commanding sweeping views of the surrounding land.

Forty-year-old Ammon Bundy exhorted others to bring their guns to Oregon to support the movement. He was one of a number of Mormons and said that he was on a mission from God. In a YouTube video he said, "I did exactly what the Lord asked me to do." Mormon Church leaders immediately condemned Bundy's actions.

The extremists employed Wise Use rhetoric, demanding that ranchers, loggers, and farmers be given control of federal lands. They did not understand the history or current use of the refuge, originally established by Theodore Roosevelt 108 years earlier. In 1935 the government bought 64,717 additional acres from Swift and Company, which had found the land unprofitable. In recent years the county residents had come together to create a land-use plan for the refuge that Interior Secretary Sally Jewell called a model. The refuge includes a complex system of canals and ditches that flow into wetlands, lakes, and ponds that provide habitat for millions of birds. Its series of canals and ditches also irrigate expanses of hay in the acreage. Local ranchers cut and bale the hay for use as winter cattle feed.[9]

Along with the Wise Use advocates and County Supremacists, the collection of insurgents at the refuge included militant gun-rights activists; avowed anti-Semites; Jon Ritzheimer, a Confederate flag–waving anti-Islamist who, saying he was willing to die, filmed a tearful good-bye to his family; Pete Santilli, an online radio host who had rallied militiamen in the 2014 Bundy confrontation; and members of disparate factions of the antigovernment Patriot movement. Brought together originally at the Bundy ranch standoff in 2014, the various ideologues found common cause in their intense resentment of the federal government.[10]

In Washington, DC, Republican congressmen voiced support for the occupiers. Kentucky representative Tom Massie said he agreed with Idaho's

Raul Labrador, who termed the action civil disobedience. Labrador criticized the media, which he said was "so quick to sort of cast aspersions on that group of people." When asked about the situation, House Speaker Paul Ryan deferred to Oregon's Greg Walden, who, although making clear that he disapproved of an armed takeover, said, "These people just want to take care of the environment—they really do. And it is the government that all too often ignores the law."[11]

Amid rumors that a Fish and Wildlife Service employee might be taken hostage, their administrator ordered staff to stay away. Although Ammon Bundy insisted their actions portended no danger for the area's citizens, the federal personnel and local officials reported harassment of their families. Residents of the nearby towns resented what they felt were intimidation and surveillance by patrolling militia. Tires were reportedly slashed, and outspoken opponent of the occupation Harney County judge Steve Grasty said he had his brake lights disconnected.

Insurgent Jon Ritzheimer was identified as one of two individuals who threatened a local woman. A criminal complaint stated: "Citizen reported to law enforcement that she heard yelling, and when she turned around, the second individual shouted 'you're BLM, you're BLM' at her. That person further stated to Citizen that they know what car she drives and would follow her home. He also stated he was going to burn Citizen's house down."[12]

Residents of the nearby town of Burns held a meeting, voting overwhelmingly for the insurgents to go home. When a group of the occupiers showed up, Judge Grasty, who was also a rancher, was one of several people who spoke directly to them, asking that they leave the county. He later commented, "Mr. Bundy walked in with his crowd. They came into a school, packing their firearms, and strategically spread around the room."[13]

The Burns Paiute Tribe was another of the local entities that protested against the insurrectionists. Members said, absent the tribe getting the land back, their priority was to protect the area's natural and cultural heritage. Tribal chairwoman Charlotte Rodrique noted the Paiutes' heartache for being forcefully removed from the land, but said that a wildlife refuge is something that benefits all. It also protects the tribe's traditional wintering grounds. "We have a good working relationship with the Malheur National Wildlife Refuge," she said. In fact, the refuge offices housed some four thousand Paiute artifacts, and others remain buried in the refuge lands. Rodrique said, "As far as I'm concerned our history is just another hostage." Tribal anger and sadness accompanied the news that the militants had built a road across the archaeologically rich ground and dug pits as repositories for garbage and human waste.[14]

In responding to tribal discontent, Ryan Bundy demonstrated limited empathy while mistaking 1950s land policy for that of the twenty-first century. He posited that in land use, "cattle ranchers and loggers should have priority" over other interests. "We recognize that the Native Americans had claim to the land, but they lost that claim," he said. "There are things to learn from cultures of the past, but the current culture is the most important."[15]

The FBI joined state and local law enforcement, using concrete ballasts to block their command center at the Burns courthouse, but they exercised restraint. They allowed the insurgents to come and go and tried to talk them into leaving the refuge but otherwise kept their distance.

Finally, on January 26, after twenty-five days, the standoff came to a head. As a group of the insurgents traveled in two vehicles toward a meeting in a neighboring county, they were pulled over by state police. One of the extremist's vehicles, an extended-cab pickup carrying five people, paused for several minutes, during which time FBI agents used bullhorns, ordering them to surrender. One of the passengers, Ryan Payne, left the vehicle and was arrested. The truck then took off at a high rate of speed.

Farther down the deserted highway, with a helicopter following overhead, the pickup attempted to circumvent a roadblock. It narrowly missed a federal officer as it careened into a two-foot snowbank and became stuck. The driver, Robert "LaVoy" Finicum, a fifty-four-year-old Arizonan who had insisted he would not go to jail, exited the vehicle. Under orders from state police officers, he raised his hands over his head. As he moved through the snow, though, he twice lowered his hands toward his left inside coat pocket. The second time two officers opened fire, killing him. In the pocket he had a 9mm semiautomatic pistol.

The other members of the party were arrested. In the truck were two loaded .223-caliber semiautomatic rifles and a loaded .38 special revolver. Two days later, the FBI released a video of the entire incident, taken from the helicopter, to counter stories that Finicum had been shot in cold blood.

Other individuals were arrested leaving the refuge or in their home states, with more than two dozen finally incarcerated. The arrestees were charged with felony counts of preventing officials from doing their duties. Ammon Bundy secured legal representation; the others declared they could not afford counsel and were assigned court-appointed attorneys.

On February 8 the entire Bundy saga took an unexpected turn. After the 2014 standoff at Bundy's ranch, Nevada state legislator Michele Fiore had called BLM officials "Nazi-minded bullies" and "thugs." She now announced that she would travel to assist one of the insurrectionists. "There is a Nevadan sitting in jail, and as an office holder, I will be there to demand his release. If that Nevadan can't leave Oregon, we will bring Nevada to him. Peaceful, of course." Fiore was also trying to convince Cliven Bundy to accompany her. Bundy told the *Las Vegas Review-Journal* by phone that he had not yet made up his mind. "I've been invited to go with [Fiore]. I haven't committed myself at all."

By February 10 he had decided to go to Burns, and he urged "patriots" and "militia" to join him. On his Facebook page he announced, "Time to wake up!" The following morning at Portland International Airport, one of his bodyguards waited at the gate, but Bundy never made it. As he debarked, FBI agents confronted the Nevadan and took him into custody.[16]

The arrest had to do with the 2014 incident at his ranch. Within the week a federal grand jury had charged Cliven, Ammon, and Ryan Bundy; Ryan Payne; and talk show host Pete Santilli with sixteen felonies. Charges included conspiracy to commit an offense against the United States; multiple counts of threatening, obstructing, impeding, and interfering with federal officers or the administration of justice; three counts of extortion involving interstate commerce; two counts of assault on a federal officer; and four counts of using firearms in relation to a crime of violence.

If convicted, prosecutors say at least $3 million in property could be forfeited. Some of the charges carry a penalty of up to five years in prison, others up to twenty years. Fines for some of the offenses can be $250,000. The U.S. attorney for Nevada, Daniel Bogden, commented, "The rule of law has been reaffirmed with these charges. Persons who use force and violence against federal law enforcement officers . . . will be brought to justice."[17]

The open rebellion of antigovernment ideologues has compounded difficulties for federal agencies on the ground. At the administrative level, the agencies have struggled to enact directives subject to dramatic changes in executive-branch philosophies.

14

Opaque Governance

Ronald Reagan's appointees overturned their predecessors' land and environmental policies in order to expand resource development. The George H. W. Bush administration followed Reagan's lead, weakening the EPA and clean-air regulations, before Bill Clinton again reversed direction. He filled the interior secretary post with former Arizona governor Bruce Babbitt, who had also headed the League of Conservation Voters. With Babbitt in charge, the department once more emphasized environmental protections.

As his presidency came to a close, Clinton worked to secure a conservation legacy. He created five national monuments, encompassing more than 1 million acres, and signed off on a "roadless rule," protecting 58.5 million acres of national forest. Wyoming District Court judge Clarence Brimmer criticized the roadless rule as having been driven "through the administrative process in a vehicle smelling of political prestidigitation." Indeed, the last-minute initiatives appeared to be bringing pressure to bear on the incoming administration's land-use options.

George W. Bush fought against that pressure. He again changed direction, immediately reintroducing the Reagan philosophy regarding land policy. After 9/11 the threat of further terrorism ensured that the government's main concern would be national security. Resource development, already a top priority, became even more critical.[1]

In seeking national office Bush had doubled down on the western image utilized by Reagan. Bush portrayed himself as a rancher-politician. He had abandoned his eastern roots and played down his Yale and Harvard Business School education. Elected to the presidency, he was shown periodically in photo ops clearing brush on his newly acquired ranch or driving foreign leaders around in an oversize four-wheel-drive pickup. When taking the country to war in Iraq, Bush used the bravado of a western actor, telling America's enemies to "bring it on." Asked about pursuing 9/11 mastermind Bin Laden, he responded, "There's an old poster out West, I recall, that says 'Wanted Dead or Alive.'"[2]

Like Reagan, Bush promoted an antiregulatory theme, denying the efficacy of the bureaucracy and trying to shrink it. In the resource arena Bush would be more successful than Reagan. There were two reasons: Bush had a Republican Congress to work with, and his Interior Department's actions were more tactical and less ostentatious. The news media had to deal with an opaqueness in his administration.[3]

Vice President Dick Cheney, who played a leading part in Bush's land policies, believed in a trustee model of government. He submitted that voters, after electing a president, needed to let him do his job his own way. Beginning the administration's second week in office, Cheney formed an energy task force that met in secret in the Executive Office Building. Over the first three and a half months of the new administration, officials from oil companies, the National Mining Association, and three dozen other trade associations met to forge a national energy policy. Environmental advocacy groups raised alarms over rumors of the closed-door meetings, but the task force excluded their input until after the policy report was substantially completed. Unsurprisingly, the proposal directed changes in economic policy that included extensive resource extraction on public lands.[4]

Like the Reagan administration, Bush's appointees seemed of a single mind. Scientists who headed agencies in the Clinton administration were replaced by appointees with law or business backgrounds. Interior Secretary Gale Norton's background included legal work for logging and mining interests working as a senior attorney at the Mountain States Legal Foundation. The agriculture secretary's chief of staff was a former lobbyist for the National Cattlemen's Beef Association. The undersecretary for natural resources had been a timber lobbyist. The Bush team worked to reverse what they perceived as obstructive environmental protections. They also sought to provide open access to federal timber, ease pollution controls, and loosen restrictions on greenhouse-gas emissions. They required a high burden of proof in order to enact environmental regulations and resisted scientists' collection of data that might meet that burden.[5]

In the dry weather year of 2001, biologists from the U.S. Fish and Wildlife Service and the National Marine Fisheries Service (NMFS) determined that water previously diverted to irrigate crops needed to remain in Upper Klamath Lake and the river below to protect the salmon population, including an endangered species, the coho. In April 2001 a judge determined that the scientific finding had to be honored. It was the first time since the Klamath Irrigation Project's inception in 1906 that fish were given priority over Klamath Basin's farm interests.

Large protests and tractor rallies by farmers that spring and summer were unsuccessful in changing the ruling. Backers of the Shovel Brigade from Elko, as well as other property rights groups, traveled to the Klamath to support the farm interests. The protesters, calling themselves the Klamath Bucket Brigade, used Wise Use strategies, including press releases decrying decisions favoring "fish over farmers" and engaging in civil disobedience—several times breaking canal headgates to release water to the farmlands. That year's agricultural losses were estimated at more than two hundred million dollars.

When the dispute had begun, Vice President Cheney instructed the Interior Department official in charge to keep him apprised. He let it be known that the administration believed farmers needed to be able to farm. In January 2002, campaigning for a Republican senator from Oregon who was in a tight race, President Bush said, "We'll do everything we can to make sure water is available for those who farm." The next day Bush's top adviser Karl Rove gave a PowerPoint presentation to fifty Interior Department administrators. Showing a slide of the Klamath's water level, Rove echoed the theme, saying the White House believed the water should be used for irrigation.[6]

With Cheney's prompting, the Interior Department contacted the National Research Council, asking that they assess the biologists' conclusion about the Klamath's water. In an interim report the NRC concluded that the finding linking river- and lake-water levels to the health of the coho salmon lacked substantiation. An NRC panel member later declared that directions to the NRC were designed to provide answers that higher-ups were looking for. He commented, "I hate it that I feel like we were manipulated for political reasons."

On March 29 Secretary of Interior Norton and Secretary of Agriculture Ann Veneman appeared at a ceremony at Klamath Falls to open Klamath River canal gates, delivering water to fifteen hundred farms. Norton said, "Our goals are to protect farm families, restore the health of the ecosystem, honor our trust responsibilities to tribes and recover endangered species." Only the first goal was met.

Members of the Yurok and Karuk Tribes, who depend on the river's salmon runs, warned administrators of danger to the fish as downstream flow levels dropped. NMFS biologists Jim Lecky and Mike Kelly, assigned to write a new opinion on the coho, requested more water for the river, but they were told to "stay consistent" with the NRC interim report. Believing he was being pressured to adopt a plan that was scientifically unsound, Kelly refused to sign the document.[7]

The administration's action, diverting the water to the farms, produced the largest fish kill in U.S. history. By the end of the summer, an estimated thirty-four thousand salmon—more than 30 percent of the year's run—had died. Several dozen of the fish were coho, a large number for a protected species. Walt Lara of the Yurok Tribal Council, said the fish were swimming in circles. "They bump up against your legs when you're standing in the water. These are beautiful, chrome-bright fish that are dying, not fish that are already spawned out."[8]

In the fall of 2002 representatives of the Klamath fishermen's association and Native American leaders dumped five hundred pounds of dead salmon on the Department of the Interior's doorstep. Mike Thompson, a California House member, said the fish represent "thousands of jobs, millions of dollars, and priceless resources that are being destroyed due to the administration's failures."[9]

At the end of November, with thousands of dead, bloated fish floating in the slow-moving river, fishermen and environmental groups took the issue to federal court. On November 28, 2002, Secretary Norton announced that the department would release water from Upper Klamath Lake, the river's reservoir. "This water will be released beginning today to meet tribal trust responsibilities and to support the migrating salmon during this emergency," she announced. Dan Keppen, executive director of the Klamath Water Users Association, representing the farmers, said he believed increasing the river flow was justified. "We had an extra slug of water available and we have a cushion right now." Sue Mastern, chair of the Yuroks, said, "We begged them for more water, starting in the spring. . . . It's just a sickening feeling."[10]

Bush officials would not concede that the die-off was caused by the low water level and warm water temperature their action had created. One official said, "It may turn out to be a natural phenomenon." Indian leaders responded that since no other river in the region had been similarly affected, the die-off could be directly tied to the diversion of the water.[11]

In explaining his role in the NRC decision, NMFS scientist Lecky explained that the findings he and his colleagues wrote were biological opinions. "I try to emphasize that they're not science documents, they're policy documents," he said. "If a biological opinion was a science document, on a par with those that appear in peer-reviewed journals, it would conclude that we don't have enough information to make a decision."[12]

His associate Kelly had a different perspective regarding the document. Months later, filing for whistle-blower protection, he asserted that it was "obvious that someone up the chain of command was applying a tremendous amount of pressure on Mr. Lecky." In May 2004, rather than write another opinion likely to be politicized, Kelly resigned from the NMFS. In his letter of resignation, he wrote, "I speak for many of my fellow biologists who are embarrassed and disgusted by the agency's apparent misuse of science."[13]

In 2002 Republican Party communication consultant Frank Luntz sent President Bush a lengthy document entitled "Straight Talk." In the section on the environment, Luntz instructed Bush on ways to convince the public that states and business interests could better protect the environment than federal regulatory management.

Luntz urged the president to pursue an environmental policy that would preserve the gains of the previous two decades without becoming too extreme. He advocated the idea that the administration's program should include removing needless bureaucratic meddling. "Give citizens the idea that progress is being frustrated by over-reaching government," he counseled, "and you will hit a very strong strain in the American psyche."

The memorandum listed eight suggestions on rhetoric. Its number-one idea was that the president convince the audience he and his party wanted to protect the environment. Luntz counseled, "Since many Americans believe Republicans do not care about the environment, you will never convince people to accept your ideas until you confront this suspicion and put it to rest."

The number-two item in the list was to emphasize instances where federal bureaucrats were failing at providing protections. Another proposal was to avoid giving the impression that Bush was probusiness. This point was followed by a strategic option: "If you must use the economic argument, stress that you are seeking 'a fair balance' between the environment and the economy. Be prepared to specify and quantify the jobs lost because of needless, excessive or redundant regulations."[14]

Bush followed Luntz's directions, publicly expressing concern for the environment, but compiling a record that favored resource users and development interests. After two years Christine Todd Whitman, a moderate former New Jersey governor, quit as Bush's EPA administrator. She later described meetings wherein she asked if there were facts to support the administration's policy, "and for that I was accused of disloyalty." Several public conflicts with others in the administration were the direct cause of her departure. The final issue was a dispute with Cheney, who insisted power plants be allowed to make alterations without new pollution controls. A federal court later found Cheney's decision violated the Clean Air Act.

Stephen L. Johnson, who had directed operations for a lab that did testing for tobacco and chemical companies, followed Whitman, serving as EPA administrator during Bush's second term. Johnson would be charged by a House oversight committee with weakening laws and failing to enforce existing rules. A specific concern was that both prosecutions of EPA cases and convictions were down by more than a third.[15]

Interior Secretary Norton's tenure infuriated environmentalists. In her decision-making she emphasized the input of those "who live on, work on, and love the land." She opened logging in areas previously managed as wilderness, urged land managers to speed up approval of gas drilling, reversed the ban on snowmobile use in Yellowstone National Park, and lobbied to allow oil drilling in Arctic National Refuge lands. Utilizing executive orders from Bush, the BLM approved 6,399 drilling permits in

2004, up from 1,803 in 1999. One of Norton's deputies, Julie A. MacDonald, resigned a week before she was to be questioned in a congressional hearing. A committee sought to ask her about altering scientific findings on endangered species and how oil lobbyists came to possess restricted documents.[16]

Other scandals involved BLM deputy secretary J. Steven Griles, who was cleared by Norton after being accused of twenty-five ethics violations. The offenses involved steering technology contracts to companies with which he had ties. Griles was later convicted of lying to Congress over his relationship with Jack Abramoff, imprisoned for the Native American lobbying scandal.[17]

In a third discreditable episode, Republican donor Harvey Frank Robbins, who had recently bought a Wyoming ranch, was charged with sixteen grazing and trespass violations. BLM director Kathleen Clarke directed deputies to craft a settlement. The final agreement suspended all charges, disregarding concerns of the Justice Department, the Wyoming assistant U.S. attorney, and BLM agents. The agreement included the extraordinary stipulation that any future actions against Robbins needed Clarke's personal approval as director. In 2006 Inspector General Earl Devaney testified before Congress: "Simply stated, short of a crime, anything goes at the highest levels of the Department of the Interior. Ethics failures on the part of senior Department officials—taking the form of appearances of impropriety, favoritism, and bias—have been routinely dismissed."[18]

At the end of Bush's second term, complaints against the administration listed in *Science* magazine included reducing the role of scientific experts, ignoring recommendations to tighten air-pollution standards, seeking ways to circumvent the National Environmental Policy Act, and attempting to withdraw protections of endangered species.[19]

15

Differing Values

George W. Bush, Dick Cheney, and their appointees had dramatically shifted the balance of land use during the administration's two terms. At the same time, another entity had positioned itself to influence decision-making. Charles and David Koch, of Koch Industries, were donating millions of dollars annually in pursuing their oft-stated goals of creating a freer society, with less government intervention, and a free market. Megacorporations, following their leadership, contributed hundreds of millions more to fund conservative political candidates and issues. They also funded right-wing think tanks and educational foundations.[1]

The Bush administration had placed twenty-three environmental regulations on a "hit list." Fourteen of those had been suggested by scholars at Mercatus, a George Mason University nonprofit institute. Mercatus was created in the mid-1980s by the Koch brothers, who contribute vast amounts of money for politicized research or endowed university positions. In 2012, for example, they donated more than $12 million to 163 colleges. But the institute at George Mason was a special project. By 2010 the Koch family had contributed more than $30 million to the public university, much of which was used to fund Mercatus. The Mercatus Institute's papers consistently attacked the EPA in favor of policies that would lessen regulations affecting Koch Industries. An example took place in 1997 when the EPA attempted to reduce surface ozone, caused in part by oil refineries—in which the Koch brothers invest.

Susan Dudley, an economist who became a top Mercatus official, argued that the EPA had not taken into account that pollution-free skies would cause more cases of skin cancer. The District of Columbia Circuit Court agreed, ruling two to one that there were "possible health benefits of ozone" and that the EPA had overstepped its authority. The judges in the majority had attended an all-expenses-paid legal junket to a Montana ranch, arranged by a research group funded by Koch family foundations. The judges denied their attendance had affected their ruling, but their extraordinary decision was later unanimously overruled by the Supreme Court.[2]

Koch Industries has a wide variety of subsidiary companies involved in activities such as coal, oil, transportation, building, and agriculture. The conglomerate earns some $115 billion a year. It has paid hundreds of millions in environmental fines, settlements, jury awards, and government-ordered repairs.

While expanding his businesses, Charles Koch promoted what he called "Market-Based Management," with profits needing to increase every quarter. Because environmental and safety compliance might cut into profits, regulations were often delayed or ignored. In a Texas case in 1996, a jury awarded claimants' families $296 million after a corrosive pipeline known to leak went unrepaired and the resultant explosion killed two teenagers.[3]

Among other cases, in 2000 the Kochs were fined $30 million for three hundred oil spills that resulted in three million gallons of crude oil polluting lakes and waterways. Later that year the company was charged with attempting to conceal the release of ninety-one metric tons of cancer-causing benzene from an oil refinery. The ninety-seven-count indictment carried potential fines of $350 million. A compromise worked out with Bush's attorney general John Ashcroft resulted in a guilty plea on one count and a $20 million fine. In 2009 the company was required to spend $500 million to fix environmental violations in seven states.[4]

Promoting individual freedom, the Koch brothers advocate deregulation of business and limited government. David Koch ran for vice president of the United States on the Libertarian ticket in 1980. The brothers subsequently became leading donors to the Republican Party, financing conservative candidacies at all levels of government.

They are abetted in fighting government regulations by executives in other megacorporations. During three years, beginning in 2005, ExxonMobil—which the United Kingdom's Royal Society and the U.S.-based Union of Concerned Scientists specifically identified as disseminating inaccurate information—spent about $9 million creating uncertainty about climate change. The Koch family spent nearly $25 million doing the same thing.[5]

For years the Kochs have hosted biannual donor retreats, attracting hundreds of individuals "who [write] seven-figure checks without flinching." The successful goal was to build a nationwide political infrastructure. They and fellow donors financed the Tea Party movement, and they have created conservative think tanks in all fifty states. In 2015, at a private meeting of 450 wealthy contributors at the St. Regis Monarch Beach Hotel at Dana Point, California, Charles Koch implored the crowd to contribute money to "help save our country. It can't be done without you and many, many others," he said. "So I pray that you will help us in this, I think, long-term, life or death struggle for our country."[6]

Another group associated with corporations and wielding profuse power in environmental and land issues is ALEC, the American Legislative Exchange Council. It is a tax-exempt nonprofit, with Koch Industries and Exxon representatives on its board and the support of 300 other corporate entities. ALEC defines itself as a tax-exempt educational organization. But it is largely involved in state lawmaking activities, creating legislation that promotes deregulation and other conservative causes.

ALEC charges $50 annual membership dues for government representatives. Membership has allowed elected officials and their families to attend the organization's meetings at five-star resorts on "conference scholarships." Corporate members, charged yearly dues from $7,000 to $25,000 or more, attend the same meetings, gaining access to legislative members. Dues enable corporate representatives to also assist in designing model legislation for state governments across the country. Among a wide range of issues, bills target the Environmental Protection Agency, the Clean Air Act, and public lands. In 2013 the organization produced some seventy bills blocking government support of renewable-energy projects. More than a quarter of all state legislators in the United States, 2,000 out of 7,383, are ALEC members.[7]

The midterm election in 2014 was a resounding political victory for the Republican Party. The Koch network spent more than $100 million to influence Senate and House races and twice that on other related activism. Republicans won full control of both chambers. This elevated conservatives to the chairs of all land-policy committees, enabling them to control legislative business.[8]

In 2015 the Republican leadership in the House of Representatives promoted the ALEC lands agenda. Alaska representative Don Young, a former ALEC member on the Committee on Natural Resources, sponsored the State-Run Federal Lands Act, authorizing federal land to be managed by the states. Jason Chaffetz, of Utah, sponsored the Disposal of Excess Federal Lands Act, offering designated federal lands in ten western states for "disposal by competitive sale." The resource committee's ranking Democrat, Raúl Grijalva of Arizona, objected. "We are seeing bills that begin to tear at the fiber of law on the federal public lands. These laws," he noted, "would allow state law to supersede the federal government."[9]

The U.S. Senate also took up the issue of disposal of lands in 2015 when it approved Senator Lisa Murkowski's nonbinding budget amendment permitting the divesting of federal lands. Votes on nonbinding budget amendments are used as a device to compel senators to declare their position on issues. Senate Amendment 838 proposed allowing "initiatives to sell or transfer to, or exchange with, a state or local government any federal land that is not within the boundary of a National Park, National Reserve, or National Monument." Sponsor Murkowski emphasized that the vote would not authorize any particular sale or transfer but would enable future legislation. The amendment was approved by the razor-thin margin of fifty-one to forty-nine. All fifty-one votes approving the measure were Republican, with all Democrats and three Republicans voting against it.[10]

The coordination of legislation at the state and federal levels is in large part the result of the right-wing think tanks across the nation. ALEC is an associate member of the State Policy Network, the umbrella group that sponsors the think tanks. SPN is another institute with deep ties to the Koch brothers' network of funders. In 2011 the combined revenue of SPN and its members was $83 million.

The Idaho Freedom Foundation (IFF), financed by tax-free donations from unidentified contributors, is illustrative of state think tanks with membership in SPN. Democrats have challenged their tax-free status, believing the group underreports its lobbying. "They're pretty darn active," said Idaho state senator Cherie Buckner Webb in 2013. "They're visible in every committee room I serve on. . . . I don't view them at all as nonpartisan." Regardless, the foundation pursues its goals of curtailing the size of government, deregulating market activities, and fighting what they term "the redistribution of wealth." Like its associated institutes, the IFF promotes transferring public lands to state control.[11]

Fred Birnbaum, formerly a manager for building materials manufacturer Boise-Cascade, is the vice president of IFF. He periodically writes op-eds on the opinion page of Boise's *Idaho Statesman* newspaper. In 2015, when Senator Murkowski's nonbinding amendment passed, Birnbaum had already been writing about the issue. Concerned with the economic effect of rural Idaho's diminished mining and timber harvesting, Birnbaum argued that the state needed to discontinue what he termed allowing "the federal government to squat on Western lands and smother our economy."

The topic of conveyance was common in Idaho because federal properties, much of it Great Basin lands, constitute well over 60 percent of the state's area. In a column three weeks after the Murkowski amendment, Birnbaum argued that there would be substantial economic benefits for Idaho in gaining jurisdiction. In that op-ed, as previously, he referenced a report by the Property and Environment Research Center, a conservative Montana think tank. The PERC study had been repeatedly cited by conservative legislators as well as newspaper editors and bloggers as evidence that states should be managing public lands. The study compared revenue generated on federal lands with those of trust lands managed by four western states: Idaho, Montana, Arizona, and New Mexico. It stated that between 2009 and 2013, the states generated an average of $14.51 for every dollar spent, while the federal government lost money, bringing in only $0.73 for each dollar spent.

Discounted in the report was the primary reason for the federal-state fee discrepancy: the fact that states are not responsible for firefighting and wildfire suppression on public lands. Federal fire costs in western states run into the billions of dollars. The authors of the PERC report waited until the conclusion of their paper to disclose that they did not address the cost of wildfire management, which, they realized, "presents a significant financial and environmental challenge on federal lands." But rather than explain why fire costs were included on the federal management side of the ledger while potential costs were omitted from the state side, they merely said, "Whether states could absorb or defray these costs, or whether other collaborative management alternatives might exist, is a question for future research."[12]

In Idaho in 2012 federal wildfire costs were $169 million, more than three times the state's total law-enforcement spending. A 2013 study by the Idaho Conservation League estimated that if lands were transferred to the state, annual costs for fire suppression, based on averages from the previous ten years, would total $134 million.[13]

Although Birnbaum stated that the PERC report debunked the notion that states might have to sell lands transferred from federal management, in fact the report's authors noted that one of the states in the study, Arizona, "earn[ed] most of its revenue from land sales and commercial leases."

Birnbaum also failed to report that resource user fees increase under state management. This is because whereas federal lands are managed for the public's multiple use, state trust lands are required to be managed for financial benefit. Owing to "higher lease rates and increased leasing competition," fees on state-managed lands in the study period were significantly higher than federal fees. For example, in 2013 federal grazing charges were below $2 per AUM, as they had remained for well over thirty years. In 2013 state fees ranged from Arizona's at just under $3 per AUM to Montana's at more than $11. The state rates had been driven up, in part, because the states allow conservation groups to bid against ranchers for parcels of rangeland. BLM and Forest Service management are prohibited under federal rules from allowing such bidding systems.[14]

In concluding his argument, Birnbaum stated, "At some point sportsmen and recreationists will have to decide whether to continue to side with the 'bicoastal' environmental gentry or partner with Idahoans who understand that multiple-use can work; with peaceful co-existence among hunters, anglers, back country enthusiasts and foresters."[15] By using the loaded descriptive "'bicoastal' environmental gentry," while identifying the timber industry as "foresters," Birnbaum obviously intended to evoke an emotional reaction. He also omitted from his conclusion other businesses likely to gain access by state management—including mineral, oil, and gas interests. In doing so, he portrayed expanded extractive industry access as less intrusive than it actually would be.

The same day Birnbaum prodded sportsmen to support conveyance, the newspaper's "Outdoors" columnist, Roger Phillips, offered a contrary opinion. Phillips said that opposition to transferring lands unites Orofino bear hunters and Ketchum backcountry skiers. He commented that selling lands where Idahoans "camp, fish, hunt, hike, ski, snowmobile, ride motorcycles, ATVs, mountain bikes, and horses, where they go for picnics, Sunday drives, and to pick huckleberries is just a bad idea." And using loaded language that mirrored Birnbaum's, Phillips warned politicians that they "can pander to the Tea Party crowd" to make a name for themselves, but, he continued, "sell off granny's secret huckleberry spot at your own peril, senators."[16]

In April 2015 an Idaho State Senate committee voted against continuing a three-year study regarding the transfer of federal lands to the state. There was an alternative, the Clearwater Basin Collaborative. Republican senator Mike Crapo worked with the Forest Service to convene individuals with a long list of differences over the six million acres of national forest. Success came through five years of negotiations by the former adversaries. The working group of twenty-three included local Native Americans and local officials, loggers and conservationists, ranchers and recreationists.

A criticism of the collaborative process is the time needed to come to agreement. The Clearwater Basin compromise took five years. Other problems with the approach include getting stakeholders to the table and a mediating entity able to keep the parties engaged. But Crapo and the Forest Service officials were able to diffuse the "us-versus-them" mentality. The results benefited all parties. The final agreement brought in sixteen million dollars in investments, national forest harvest levels that increased by 50 percent, the reduction of catastrophic fire risk across sixty thousand acres, the restoration of more than three thousand miles of trails, and eighty jobs being created or sustained.[17]

Conclusion

The history of public land policy shows the significance of the government's changeable disposition. It underlines the impact of responses by activists and the major political parties. It brings attention to the questions raised regarding the possible transference of lands to states or private entities.

In the second decade of the twenty-first century, the protagonists continue to confront one another. In the West there have been instances of open rebellion. In Washington, DC, arguments embroil all three branches of government. The primary issues involve who should make the decisions and whether emphasis should be afforded the economy and resource expansion or ecological concerns and western heritage.

In 2015 the College of Colorado sponsored a "State of the Rockies Project" survey. It questioned citizens of six western states that have considered challenging management of public lands: Arizona, Colorado, Montana, New Mexico, Utah, and Wyoming. Asked to list "top priorities," 82 percent of respondents said "protecting and conserving natural areas for future generations," while only 40 percent listed "making sure resources . . . are available for development."[1]

The passage that year of the Murkowski Senate amendment, enabling the transfer of federal lands, illustrated the marked inroads of the states' rights movement and political and fiscal conservatives. It is difficult to ascertain precisely what the vote represents or what it portends. It is likely that members supporting the amendment differed in their motives. Idaho's

Mike Crapo disappointed constituents he had worked with on Clearwater Basin by voting with the majority. He later told the *Idaho Statesman* that laying the groundwork for conveying lands did not mean he actually supported doing it.[2]

To transfer lands, fundamental questions must be answered. Do legislators who are seeking to reduce federal holdings want the lands conveyed to the states or sold to private interests? Will ranchers support transference to state control if their fees will be raised and bidding for rangeland is allowed? Because opinion polls show substantial majorities of Americans favor retaining public lands, will legislators and a future president act when the removal of specific properties is proposed?

Finally, it is beneficial to ask what effects conveyance might have. The federal government has allowed hikers, hunters, and fishermen free access to public lands and charged below-market fees for ranchers and other land users. States, with periodic taxpayer revolts, would not have the same ability to reject promised profits from competing sources. Past state management practices, including the widespread sale of public lands, do not inspire confidence in states' ability to maintain public access.

President Barack Obama extended access, using executive orders to circumvent a Republican Congress. His actions caused opponents to accuse him of "surreptitious land grabs." He used the Antiquities Act thirteen times and signed a bill designating two million acres as wilderness.[3]

The 2016 presidential campaign emphasized the critical importance of who serves as chief executive. Four of the five leading Republican candidates advocated the divestiture of public lands, and all attacked what they believed to be misuse of environmental laws. Candidate Ted Cruz put out a television ad in Nevada using a timeworn, politically expedient ploy. He said he would "fight night and day to return full control of Nevada's lands to its rightful owners, its citizens." Cruz's statement falsely implies that the property had once belonged to the state's residents. Similar to other western states, the Nevada Enabling Act of 1864 states: "The people inhabiting said territory do agree and declare that they forever disclaim all right and title to the unappropriated public lands lying within said territory."[4]

The eventual Republican standard-bearer and future president, Donald Trump, commented that the EPA "is a disgrace" and needed to be abolished. When asked who would protect the environment, he responded,

"We'll be fine with the environment. We can leave a little bit, but you can't destroy business."[5]

Profit fuels the U.S. economy. Business interests will always attempt to gain an advantage over competitors. In the past that has often involved abuse of the land. Having the federal government as steward has blunted some of the damage. Transference or allowing free-market access to public lands would create opportunities for profitability and boosts to local employment. But weighing the immediacy of economic considerations alongside environmental ethics and cultural history underscores the reality that development would change the West's demographics, landscape, and identity.

Federal management has been uneven, unprofitable, and in arenas ineffective. It has been obtrusive and unfair to some users. Yet it has sustained a western lifestyle for many others. While facilitating a wide variety of activities, the federal government has preserved parks, forests, rangelands, deserts, wilderness, and the region's ecological heritage.

Unbounded, undeveloped expanses, embodying America's frontier legacy, differentiate the West from the rest of the country. This fact should be kept in mind when deciding the fate of its public lands.

Notes

Introduction

1. Christopher Ketcham, "The Great Republican Land Heist: Cliven Bundy and the Politicians Who Are Plundering the West," *Harper's Magazine*, Feb. 2015, 24.
2. *United States v. Cliven Bundy*, U.S. District Court District of Nevada, Criminal Complaint Case No. 2:16-mj-00127-PAL (Feb. 11, 2016); "Armed Protesters Head to Nevada in Fight over Cattle Ranch," CBS/AP, April 13, 2014, www.cbsnews.com; Dylan Scott, "Why Bundy Ranch Thinks America's Sheriffs Can Disarm the Feds," *Talking Points Memo*, April 15, 2014, http://talkingpointsmemo.com/.
3. "Hannity," *Fox News*, April 15, 16, 2014; "Cliven Bundy on Blacks: 'Are They Better Off as Slaves?,' " *Washington Post*, April 24, 2014.
4. Ketcham, "Great Republican Land Heist," 24; "Nevada Politicians Fan the Flames of Bundy-BLM Clash," *Las Vegas Review-Journal*, April 19, 2014; "Nevada Officials Blast Feds over Treatment of Cattle Rancher Cliven Bundy," Fox News, April 10, 2014; "Armed Protesters Head to Nevada"; "SPLC Report: Bundy Ranch Standoff Was Highly Coordinated, Reflecting Threat of Larger Far-Right Militia Movement," Southern Poverty Law Center, July 10, 2014, www.splcenter.org; "A Much Larger and More Dangerous Movement," *Salon*, July 22, 2014, www.salon.com.
5. "S.Amdt. 838 to S.Con.Res.11," 114th Cong., March 25, 2015, www.congress.gov.
6. The West's untapped natural resources and its history as a symbol of American values caused historian Howard Lamar to describe it as a "persistent frontier." See Howard R. Lamar, "Persistent Frontier: The West in the Twentieth Century," *Western Historical Quarterly* 4, no. 1 (1973): 4–25.
7. "Federal Land Ownership: Overview and Data," Congressional Research Service, Feb. 8, 2012, https://fas.org.
8. A further deterioration of traditional commodity developers' influence occurred with the gentrification of rural expanses. Ranches proximate to cities were sold and subdivided for housing tracks. In scenic areas surrounding towns like Vail, Park City, Jackson Hole, Taos, and Bozeman, a postindustrial middle class took up residence. Tensions developed as, rather than the productive potential of the land, the new inhabitants sought experiences offered by the West's natural beauty. Should scarce water be used for recreation or irrigation? Should wolves be treated as part of the ecosystem or a dire

threat? See J. Dwight Hines, "On Water and Wolves: Toward an Integrative Political Ecology of the 'New' West," in *Bridging the Distance: Common Issues of the Rural West*, edited by David B. Danbom (Salt Lake City: University of Utah Press, 2015), 83–100.

9. The title of this book uses the plural of the descriptive to generally categorize clashes as an ongoing series of events involving the Far West.

10. James Morton Turner, "The Specter of Environmentalism: Wilderness, Environmental Politics, and the Evolution of the New Right," *Journal of American History* 96, no. 1 (2009): 145–48; "2015 National Environmental Scorecard, 1st Session of the 114th Congress," League of Conservation Voters, http://scorecard.lcv.org.

11. "Nevada Statewide Policy Plan for Public Lands," Nevada Division of State Lands and Nevada Cities and Counties, 1985, http://lands.nv.gov; "Report Shows Idaho Has Sold 41 Percent of Its Land since Statehood," *Idaho Statesman* (Boise), May 4, 2016; Holly Fretwell and Shawn Regan, "Divided Lands: State vs. Federal Management in the West," Property and Environment Research Center, 2015, www.perc.org.

12. "Summary: Wildfire Costs, New Development, and Rising Temperatures," Headwaters Economics, April 2016, http://headwaterseconomics.org.

13. Steve H. Hanke, "Why the Feds (Still) Own So Much of the Country," Foundation for Economic Education, Jan. 11, 2016, https://fee.org; Ted Cruz, *A Time for Truth: Reigniting the Promise of America* (New York: Broadside Books, 2015), 197.

1. Creating the Federal Domain

1. Frederick D. Drake and Lynn R. Nelson, eds., *States' Rights and American Federalism: A Documentary History* (Westport, CT: Greenwood Press, 1999), viii–ix; Ralph Ketcham, ed., *The Anti-Federalist Papers and the Constitutional Convention Debates: The Clashes and the Compromises That Gave Birth to Our Form of Government* (1988; reprint, New York: New American Library, 2003), 15–19.

2. "Jackson's Proclamation to the People of South Carolina, Dec. 10, 1832," in *Documents of American History*, edited by Henry Steele Commager (New York: Appleton-Century-Crofts, 1963), 1:261–69.

3. Roosevelt to Henry Joseph Haskell, Dec. 28, 1918, quoted in Lewis L. Gould, *Grand Old Party: A History of the Republicans* (New York: Random House, 2003), 193–94.

4. Samuel Trask Dana and Myron Krueger, *California Lands: Ownership, Use, and Management* (Washington, DC: American Forestry Association, 1958), 42–43; David Igler, "Engineering the Elephant: Industrialism and the Environment in the Greater West," in *A Companion to the American West*, edited by William Deverell (Malden, MA: Blackwell, 2007), 103; Bernard DeVoto, *The Western Paradox: A Conservation Reader*, edited by Douglas Brinkley and Patricia Nelson Limerick (New Haven, CT: Yale University Press, 2001), 135. This book includes a number of DeVoto's *Harper's Magazine* articles. His discussion of grazing on public lands was in "The Sturdy Corporate Homesteader," which appeared in *Harper's*, May 1953. See also Paul W. Gates, *History of Public Land Law Development* (Washington, DC: Wm. W. Gaunt & Sons, 1987), 575.

5. DeVoto, *Western Paradox*, 136; Doris Kearns Goodwin, *The Bully Pulpit: Theodore Roosevelt, William Howard Taft, and the Golden Age of Journalism* (New York: Simon & Schuster, 2013), 585; Edmund Morris, *The Rise of Theodore Roosevelt* (New York: Random House, 1979), xiv. Critics protested that Roosevelt did not go far enough, that in fighting the monopolies he tempered his actions to placate the magnates' purposes. Roosevelt countered by arguing that change came about "by the actions of men who

take the next step; not those who theorize about the 200th step." His example was that "'it was Lincoln,' not Wendell Phillips, who 'saved the Union and abolished slavery'" (Goodwin, *Bully Pulpit*, 540). For the classic robber-baron argument against his intent and the efficacy of his attacks on the trust movement, see Gustavus Myers, *History of the Great American Fortunes* (1907; reprint, New York: Random House, 1964), 592–94.
6. Albert F. Potter quoted in William D. Rowley, *U.S. Forest Service Grazing and Rangelands: A History* (College Station: Texas A&M University Press, 1985), 16.
7. Char Miller, *Gifford Pinchot and the Making of Modern Environmentalism* (Washington, DC: Island Press, 2001), 181.
8. *Stearns v. Minnesota*, 179 U.S. 223, www.law.cornell.edu.
9. Miller, *Gifford Pinchot*, 147.
10. Stuart L. Udall, *The Quiet Crisis* (New York: Holt, Rinehart, and Winston, 1963), 97–98.
11. "Letter to Denver Republican," *Aspen (CO) Democrat*, March 14, 1907; "The Land Convention," *Steamboat (CO) Pilot*, May 22, 1907; *Castle Rock (CO) Journal*, May 24, 1907.
12. Miller, *Gifford Pinchot*, 158–59.
13. Char Miller, *Public Lands, Public Debates: A Century of Controversy* (Corvallis: Oregon State University Press, 2012), 19, 26–30; "Senator Telluride Talks Back," *Aspen (CO) Daily Times*, Aug. 11, 1906; G. Michael McCarthy, "The First Sagebrush Rebellion: Forest Reserves and States Rights in Colorado and the West, 1901–1907," in *The Origins of the National Forests: A Centennial Symposium, 1992*, edited by Harold K. Steen (Durham, NC: Forest History Society, 1992), www.foresthistory.org.
14. G. Michael McCarthy, "The Pharisee Spirit: Gifford Pinchot in Colorado," *Pennsylvania Magazine of History and Biography* 97, no. 3 (1973): 364–68.
15. Leisl Carr Childers, *The Size of the Risk: Histories of Multiple Use in the Great Basin* (Norman: University of Oklahoma Press, 2015), 18; "The Farm and Range," *Akron (OH) Weekly Pioneer Press*, Dec. 8, 1905.
16. "It Is Illegal," *Steamboat (CO) Pilot*, Nov. 14, 1906; *Castle Rock (CO) Journal*, May 24, 1907.
17. Gates, *History of Public Land Law Development*, 582; Miller, *Gifford Pinchot*, 163–64.
18. "The Conservation of Natural Resources: Roosevelt's Seventh Annual Message to Congress," Dec. 3, 1907, in *Documents of American History*, edited by Commager, 49.
19. *Proceedings of a Conference of Governors, May 13–15, 1908* (Washington, DC: Government Printing Office, 1909), 129, 144, 348–49; "Declaration of the Conservation Conference," May 15, 1908, in *Documents of American History*, edited by Commager, 50–52.

2. Interior Battles

1. John T. Ganoe, "Proceedings of the National Irrigation Congress (1909)," *Pacific Historical Review* 3, no. 3 (1934): 323; Timothy Egan, *The Big Burn: Teddy Roosevelt & the Fire That Saved America* (Boston: Houghton Mifflin Harcourt, 2009), 21; William D. Rowley, *The Bureau of Reclamation: Origins and Growth to 1945* (Washington, DC: Government Printing Office, 2004), 1:159–60.
2. Stuart L. Udall, *The Quiet Crisis* (New York: Holt, Rinehart, and Winston, 1963), 132.
3. "Ballinger Reverses Policy of Garfield," *Los Angeles Herald*, May 5, 1909; "Ballinger Will Bear Watching," *San Francisco Call*, May 9, 1909.
4. Douglas H. Strong, *Tahoe: From Timber Barons to Ecologists* (1984; reprint, Lincoln: University of Nebraska Press, 1999), 54; "The Tahoe and Hetch Hetchy Controversies," *San Francisco Call*, Oct. 11, 1909, "Welch Bill to Give City a Square Deal," *San Francisco*

Call, Feb. 4, 1911, "Auburn Protests the Lake Tahoe Contract," *San Francisco Call,* Feb. 5, 1911.

5. George Hinkle and Bliss Hinkle, *Sierra-Nevada Lakes* (1949; reprint, Reno: University of Nevada Press, 1984), 338–40; Michael J. Makley, *Saving Lake Tahoe: An Environmental History of a National Treasure* (Reno: University of Nevada Press, 2013), 28–29; "Use of Tahoe's Waters Opposed," *San Francisco Call,* Feb. 1, 1911; "Welch Bill to Give City a Square Deal."
6. Makley, *Saving Lake Tahoe,* 29.
7. Rachel White Scheuering, *Shapers of the Great Debate on Conservation: A Biographical Dictionary* (Westport, CT: Greenwood Press, 2004), 15–23; James L. Penick Jr., "The Age of the Bureaucrat: Another View of the Ballinger-Pinchot Controversy," *Forest History Newsletter* 7, nos. 1–2 (1963): 16, 21. For a specific example of Ballinger's policy where he reduced protected lands from more than a million acres to fewer than 150,000, see "Explains Policy as to the Power Sites Lands," *San Francisco Call,* July 17, 1909.
8. Scheuering, *Shapers of the Great Debate on Conservation,* 23.
9. "Taft Upholds Ballinger," *San Francisco Call,* Sept. 16, 1909; Doris Kearns Goodwin, *The Bully Pulpit: Theodore Roosevelt, William Howard Taft, and the Golden Age of Journalism* (New York: Simon & Schuster, 2013), 614–15.
10. "Congress Will Probe Quarrel," *Los Angeles Herald,* Jan. 6, 1910, "Wickersham Is with Ballinger," *Los Angeles Herald,* Jan. 7, 1910, "Taft Ousts Pinchot," *Los Angeles Herald,* Jan. 8, 1910.
11. Goodwin, *Bully Pulpit,* 622–26; "Taft Says He Told Lawler to Write Ballinger Letter," *Los Angeles Herald,* May 16, 1910; "Ballinger Flays, Then Drops Kirby," *Los Angeles Herald,* May 17, 1910.
12. "Big Stick Swings in the Fight to Crush Ballinger," *Los Angeles Herald,* July 6, 1910; "The Expected Whitewash in Ballinger Case," *San Francisco Call,* May 30, 1910.
13. "Repudiation of Ballinger and Victory for Pinchot," *Washington Times,* June 26, 1911.
14. Penick, "Age of the Bureaucrat," 18–19, 21.
15. Lewis L. Gould, *Grand Old Party: A History of the Republicans* (New York: Random House, 2003), 235–36.
16. Calvin Coolidge, Sixth Annual Message, Dec. 4, 1928, American Presidency Project, www.presidency.ucsb.edu; Michael McGerr, *A Fierce Discontent: The Rise and Fall of the Progressive Movement in America* (Oxford: Oxford University Press, 2003), 312.
17. Jacqueline Vaughn Switzer, *Green Backlash: The History and Politics of Environmental Opposition in the U.S.* (Boulder, CO: Lynne Rienner, 1997), 41; William D. Rowley, *U.S. Forest Service Grazing and Rangelands: A History* (College Station: Texas A&M University Press, 1985), 144.
18. Paul W. Gates, *History of Public Land Law Development* (Washington, DC: Wm. W. Gaunt & Sons, 1987), 524–28; "Utah Governor George H. Dern Responds to the Public Lands Debate, 1932," in *Major Problems in the History of the American West: Documents and Essays,* edited by Clyde A. Milner II, Anne M. Butler, and David Rich Lewis, 2nd ed. (Boston: Wadsworth Cengage Learning, 1997), 489.
19. "The Taylor Act," June 28, 1934, in *Documents of American History,* edited by Henry Steele Commager (New York: Appleton-Century-Crofts, 1963), 2:294–96.

3. Rangeland Battles

1. Peter Woll, *American Bureaucracy* (New York: W. W. Norton, 1963), 4, 28, 56.

2. Ibid., 49, 60; Karen R. Merrill, *Public Lands and Political Meaning: Ranchers, the Government, and the Property between Them* (Berkeley: University of California Press, 2002), 139. As to Ickes managing both the Public Works Administration and the Interior Department, years later the president commented that there had been no previous interior secretaries who could have filled both jobs at the same time. Harold L. Ickes, *The Secret Diary of Harold L. Ickes* (New York: Simon and Schuster, 1954), 2:674.
3. Merrill, *Public Lands and Political Meaning*, 140; Ickes, *Secret Diary*, 1:143, 172; Jacqueline Vaughn Switzer, *Green Backlash: The History and Politics of Environmental Opposition in the U.S.* (Boulder, CO: Lynne Rienner, 1997), 53.
4. For discussions of weaknesses in the Taylor Grazing Act, see Paul J. Culhane, *Public Lands Politics: Interest Group Influence on the Forest Service and the Bureau of Land Management* (New York: Resources for the Future, 1981), 88–89; Donald J. Pisani, "The Many Faces of Conservation," in *Taking Stock: American Government in the Twentieth Century*, edited by Morton Keller and R. Shep Melnick (Cambridge: Cambridge University Press, 1999), 146–47; and Merrill, *Public Lands and Political Meaning*, 141–42.
5. Farrington R. Carpenter, "Historical Interview," interviewed by Jerry A. O'Callahan, Department of the Interior, distributed via BLM Information Memorandum No. 81-229, July 9, 1981, www.rangebiome.org; Phillip O. Foss, "The Determination of Grazing Fees on Federally-Owned Land," *Journal of Farm Economics* 41, no. 3 (1959): 537; Merrill, *Public Lands and Political Meaning*, 148, 169.
6. Carpenter, "Historical Interview."
7. American Presidency Project, www.presidency.ucsb.edu.
8. Foss, "Determination of Grazing Fees," 537–38.
9. Ickes, *Secret Diary*, 1:101. Carpenter later said that he refused to resign until Ickes wrote a letter saying Carpenter's work had been satisfactory. Although several days later Ickes provided him with the letter, Ickes's initial response had been, "That would be a 'hell of a long time'" (Merrill, *Public Lands and Political Meaning*, 248n4).
10. Char Miller, *Gifford Pinchot and the Making of Modern Environmentalism* (Washington, DC: Island Press, 2001), 347–48; Hal K. Rothman, "A Regular Ding-Dong Fight: Agency Culture and Evolution in the NPS-USFS Dispute, 1916–1937," *Western Historical Quarterly* 20, no. 2 (1989): 159; Miller, *Gifford Pinchot*, 353–55.
11. William D. Rowley, *U.S. Forest Service Grazing and Rangelands: A History* (College Station: Texas A&M University Press, 1985), 173–74. For information on McCarran's role in Nevada politics, see Jerome E. Edwards, *Pat McCarran: Political Boss of Nevada* (Reno: University of Nevada Press, 1982).
12. Culhane, *Public Lands Politics*, 87; Merrill, *Public Lands and Political Meaning*, 182; Glen D. Weaver, "Nevada's Federal Lands," *Annals of the Association of American Geographers* 59, no. 1 (1969): 35.
13. "Taylor Grazing Act," Bureau of Land Management, Casper Field Office, Jan. 13, 2011, www.publiclandsranching.org; Paul W. Gates, *History of Public Land Law Development* (Washington, DC: Wm. W. Gaunt & Sons, 1987), 620; Foss, "Determination of Grazing Fees," 538; Merrill, *Public Lands and Political Meaning*, 184–85.
14. Thomas L. Fleisher, "Land Held Hostage: A History of Livestock and Politics," www.publiclandsranching.org; Merrill, *Public Lands and Political Meaning*, 180.
15. Foss, "Determination of Grazing Fees," 541, 543.
16. Ibid., 542–43.
17. James R. Skillen, *The Nation's Largest Landlord: The Bureau of Land Management in the*

American West (Lawrence: University Press of Kansas, 2009), 36–37; Fleisher, "Land Held Hostage"; Culhane, Public Lands Politics, 89.
18. Gates, History of Public Land Law Development, 623; Switzer, Green Backlash, 42; Culhane, Public Lands Politics, 89; Merrill, Public Lands and Political Meaning, 198–99.
19. Tom Wolf, Arthur Carhart: Wilderness Prophet (Boulder: University Press of Colorado, 2008), 212, 227–28, 280.
20. Bernard DeVoto, The Western Paradox: A Conservation Reader, edited by Douglas Brinkley and Patricia Nelson Limerick (New Haven, CT: Yale University Press, 2001), 46–50, 65–67. This book includes a number of DeVoto's Harper's Magazine articles. This is from "The West against Itself," Harper's, Jan. 1947.
21. Switzer, Green Backlash, 42; Merrill, Public Lands and Political Meaning, 180–81, 198–99; Culhane, Public Lands Politics, 89. When DeVoto wrote an editorial in his "Easy Chair" column attacking the Federal Bureau of Investigation (FBI), J. Edgar Hoover began an investigation of DeVoto. J. Elmer Brock responded to an inquiry, "More power to you in your attack on DeVoto. As far as the sentiments of the stockmen in the West, we would like to see Bernard hanging from a cottonwood limb rather than reclining in his 'easy chair.'" Tom Knudson, High Country News, Aug. 8, 1994, www.hcn.org.
22. William L. Graf, Wilderness Preservation and the Sagebrush Rebellions (Savage, MD: Rowman & Littlefield, 1990), 174; Gates, History of Public Land Law Development, 629; DeVoto, Western Paradox, 130–31, from "Billion Dollar Jackpot," Harper's, Feb. 1953. As for the principals, in 1948 DeVoto had won the Pulitzer Prize for History and the Bancroft Prize for his book Across the Wide Missouri. He continued writing for Harper's and later won the National Book Award for The Course of Empire. McCarran, in his third term as senator, had health problems, suffering a heart attack in December 1946 and others in 1947 and 1951. He continued in office, turning his attention to attempting to stop postwar refugees from immigrating. Fearful that communism was threatening the American way of life, he worried over "red" newspapers and "commie" plants. In 1950 he authored the McCarran Internal Security Act, requiring the registration of all communists and communist-supported organizations. Over the course of his long government service, McCarran's popularity waned. In 1934 Time had adjudged him one of the most popular members of the Senate: "He is considered intellectually honest, frank, logical, and has a way of coming to the point without a smokescreen of oratory." By 1950 Time listed him as one of ten expendable senators, calling him "pompous, vindictive, and power grabbing." He died still in office in 1954. DeVoto, Western Paradox, 74, 113; Edwards, Pat McCarran, 132, 193–95; John Whiteclay Chambers II, ed., The Oxford Companion to American Military History (Oxford: Oxford University Press, 1999), 332.

4. Multiple Use

1. Although their biggest supporter, Senator McCarran actually hurt the interests of some Nevada ranchers by lobbying for the establishment of a military training base on Nevada's public land. The Tonopah Bombing Range, the first of a number of installations—including the nuclear test range in the southern part of the state—was an economic boon. But it forced certain ranchers to make "satisfactory adjustments." Several ranchers complained to McCarran that the army was taking their grazing land just as the government had stolen it from the Shoshone Indians. One rancher quipped that the Shoshone had starved, and, "I guess we will, too." The adjustments ranchers were required to make later included the effects of nuclear fallout on humans and animals. Leisl Carr Childers,

"The Angry West: Understanding the Sagebrush Rebellion in Rural Nevada," in *Bridging the Distance: Common Issues of the Rural West*, edited by David B. Danbom (Salt Lake City: University of Utah Press, 2015), 222–23, 225. For an extensive discussion of nuclear testing affecting Nevada ranchers, see Leisl Carr Childers, *The Size of the Risk: Histories of Multiple Use in the Great Basin* (Norman: University of Oklahoma Press, 2015), chaps. 3, 4. Similar to Nevada's nuclear testing, Utah hosted the Dugway Proving Ground, which tested chemical and biological agents and where, in 1968, more than six thousand sheep died when nerve gas drifted into grazing lands. Danbom, introduction to *Bridging the Distance*, 3.

2. John Muir, "The Range of Light," in *A Treasury of the Sierra Nevada*, edited by Robert Leonard Reid (Berkeley, CA: Wilderness Press, 1988), 169; Michael P. Cohen, *The History of the Sierra Club, 1892–1970* (San Francisco: Sierra Club Books, 1988), 17.

3. Dan Flores, "Societies to Match the Scenery: Twentieth-Century Environmental History in the American West," in *A Companion to the American West*, edited by William Deverell (2004; reprint, Malden, MA: Blackwell, 2007), 265; R. McGreggor Cawley, *Federal Land, Western Anger: The Sagebrush Rebellion and Environmental Politics* (Lawrence: University Press of Kansas, 1993), 18–19; Rob Chaney, "Brandborg Recalls Effort to Pass Wilderness Act 50 Years Ago," *Missoula (MT) Missoulian*, Sept. 1, 2014, http://missoulian.com.

4. "David Brower Timeline," Sierra Club, http://content.sierraclub.org. The Sierra Club had 7,000 members in 1950, doubled its membership by 1960, doubled again by 1965, and tripled by 1970. Adam Rome, "Give Earth a Chance: The Environmental Movement and the Sixties," *Journal of American History* 90, no. 2 (2003): 525–54; William L. Graf, *Wilderness Preservation and the Sagebrush Rebellions* (Savage, MD: Rowman & Littlefield, 1990), 242. The Sierra Club website time line lists the organization's 1970 membership at more than 114,000. http://vault.sierraclub.org.

5. Lewis L. Gould, *Grand Old Party: A History of the Republicans* (New York: Random House, 2003), 334; Barry Goldwater, *The Conscience of a Conservative* (1960; reprint, Washington, DC: Regnery, 1990), 3.

6. Michael McGerr, "Is There a Twentieth Century West?," in *Under an Open Sky: Rethinking America's Past*, edited by William Cronon, George Miles, and Jay Gitlin (New York: W. W. Norton, 1992), 251; Peter Iverson, *Barry Goldwater: Native Arizonan* (Norman: University of Oklahoma Press, 1997), 131; Goldwater, *Conscience of a Conservative*, 6–8.

7. Goldwater, *Conscience of a Conservative*, 9–10; Barry M. Goldwater, *With No Apologies: The Personal and Political Memoirs of United States Senator Barry M. Goldwater* (New York: William Morrow, 1979), 76.

8. Richard White, *It's Your Misfortune and None of My Own: A New History of the American West* (Norman: University of Oklahoma Press, 1991), 602–3, 609; William D. Rowley, *The Bureau of Reclamation: Origins and Growth to 1945* (Washington, DC: Government Printing Office, 2004), 1:9–10. As an example of defense expenditures in Arizona with Senator Goldwater representing the state, in 1963 it received $286 million in defense contracts and an estimated $130 million in payroll. Iverson, *Barry Goldwater: Native Arizonan*, 193–200.

9. Charles A. Reich, "The Public and the Nation's Forests," *California Law Review* 50, no. 3 (1962): 383; Cawley, *Federal Land, Western Anger*, 19. The Forest Service had begun providing for outdoor recreation many years earlier when timber demands lessened during the Great Depression and the agency was threatened with being subsumed into

the Department of the Interior. Richard Freeman, "The EcoFactory: The United States Forest Service and the Political Construction of Ecosystem Management," *Environmental History* 7, no. 4 (2002): 633, 649n7.
10. Cawley, *Federal Land, Western Anger*, 19–20.
11. Carl Abbott, "The Urban West and the Twenty-First Century," in *Major Problems in the History of the American West: Documents and Essays*, edited by Clyde A. Milner II, Anne M. Butler, and David Rich Lewis, 2nd ed. (Boston: Wadsworth Cengage Learning, 1997), 478.
12. James R. Skillen, *The Nation's Largest Landlord: The Bureau of Land Management in the American West* (Lawrence: University Press of Kansas, 2009), 44–45; Childers, *Size of the Risk*, 124; Jedediah Smart Rodgers, *Roads in the Wilderness: Conflict in Canyon Country* (Salt Lake City: University of Utah Press, 2013), 66–67.
13. Stuart L. Udall, *The Quiet Crisis* (New York: Holt, Rinehart, and Winston, 1963), xiii.
14. Flores, "Societies to Match the Scenery," 265; Udall, *The Quiet Crisis*, 178.
15. Skillen, *Nation's Largest Landlord*, 57; Cawley, *Federal Land, Western Anger*, 21, 75; Thomas M. Quigley, R. Garth Taylor, and R. McGreggor Cawley, "Public Resource Pricing: An Analysis of Range Policy," U.S. Forest Service, Pacific Northwest Research Station, PNW-RS-158, Aug. 1988, www.researchgate.net.
16. Wallace Stegner, "The Geography of Hope," printed by Stanford University, Eco Speak, http://web.stanford.edu.
17. Wallace Stegner, "The Wilderness Idea," in *Voices for the Wilderness*, edited by William Schwartz (New York: Ballantine Books, 1969), 289.
18. Curt Meine, *Aldo Leopold: His Life and Work* (Madison: University of Wisconsin Press, 2010), 178; Aldo Leopold, *A Sand County Almanac: With Other Essays on Conservation from Round River* (New York: Random House, 1970), 239.
19. Chaney, "Brandborg Recalls Effort"; Howard Zahniser, "Wilderness Forever," in *Voices for the Wilderness*, edited by Schwartz, 105–6; James Morton Turner, "The Specter of Environmentalism: Wilderness, Environmental Politics, and the Evolution of the New Right," *Journal of American History* 96, no. 1 (2009): 126; Tom Wolf, *Arthur Carhart: Wilderness Prophet* (Boulder: University Press of Colorado, 2008), 257; Cawley, *Federal Land, Western Anger*, 26–27.
20. Turner, "Specter of Environmentalism," 127.
21. Chaney, "Brandborg Recalls Effort."
22. Turner, "Specter of Environmentalism," 128; Cawley, *Federal Land, Western Anger*, 24–26. Environmentalists railed against Aspinall for protecting extractive industries in blocking passage of the Redwoods National Park bill for a half decade. The bill finally passed in 1968.
23. Brian Allen Drake, "The Skeptical Environmentalist: Senator Barry Goldwater and the Environmental Management State," *Environmental History* 15, no. 4 (2010): 587–96; Iverson, *Barry Goldwater: Native Arizonan*, 133.
24. Rome, "Give Earth a Chance," 527–29, 536–37, 547–49.
25. Cawley, *Federal Land, Western Anger*, 43.
26. Chaney, "Brandborg Recalls Effort"; Turner, "Specter of Environmentalism," 128.
27. Udall, *The Quiet Crisis*, 184. One of Udall's major accomplishments was freezing disposition of Alaska state land claims in 1966 pending settlement of the state's Native peoples' claims. His action led to the 1971 Alaska Native Claims Settlement Act, allowing Natives to gain forty-four million acres of traditional Alaskan lands as well as nearly one billion

dollars. Stephen Haycox, *Alaska: An American Colony* (Seattle: University of Washington Press, 2002), 279.
28. Lyndon B. Johnson, "Special Message to the Congress on Conservation and Restoration of Natural Beauty," Feb. 8, 1965, American Presidency Project, www.presidency.ucsb.edu; Cawley, *Federal Land, Western Anger*, 27–28.
29. Rome, "Give Earth a Chance," 532–34; Cawley, *Federal Land, Western Anger*, 29.

5. The Radical New Conservation

1. R. McGreggor Cawley, *Federal Land, Western Anger: The Sagebrush Rebellion and Environmental Politics* (Lawrence: University Press of Kansas, 1993), 17–19.
2. Steven F. Hayward, "From Goldwater to the Tea Party," Federalist, Sept. 23, 2014, http://thefederalist.com; Peter Iverson, *Barry Goldwater: Native Arizonan* (Norman: University of Oklahoma Press, 1997), xiii–xiv.
3. Western historians William Cronon, George Miles, and Jay Gitlin said, "If the presidency of Ronald Reagan taught us nothing else, it surely affirmed the continuing power of western symbolism to express the identity and vision of ordinary Americans." William Cronon, George Miles, and Jay Gitlin, eds., *Under an Open Sky: Rethinking America's Past* (New York: W. W. Norton, 1992), 5.
4. Nick Perlstein, *Nixonland: The Rise of a President and the Fracturing of America* (New York: Scribner, 2008), 83, 90–91.
5. Russell Train, "The Environmental Record of the Nixon Administration," *Presidential Studies Quarterly* 26, no. 1 (1996): 185.
6. For a specific example of industry using a lobbyist, see Jacqueline Vaughn Switzer, *Green Backlash: The History and Politics of Environmental Opposition in the U.S.* (Boulder, CO: Lynne Rienner, 1997), 109.
7. Train, "Environmental Record of the Nixon Administration," 189–91.
8. Ibid., 189, 195.
9. Cawley, *Federal Land, Western Anger*, 62–65. For an example of an illegal ORV demonstration that led to criminal charges, see chapter 13.
10. James R. Skillen, *The Nation's Largest Landlord: The Bureau of Land Management in the American West* (Lawrence: University Press of Kansas, 2009), 93–94.
11. Ibid., 95–96.
12. Cawley, *Federal Lands, Western Anger*, 136–37. Because Nixon's plan to privatize lands was designed to acquire new lands, it differs significantly from Hoover's idea of transference to the states and a later proposal by the Reagan administration to privatize lands to reduce the national debt. Cawley discusses privatization and his conclusions regarding the Reagan administration's efforts on 137–42.
13. Bruce J. Schulman, *The Seventies: The Great Shift in American Culture, Society, and Politics* (New York: Da Capo Press, 2002), 30; Skillen, *Nation's Largest Landlord*, 112; Samuel P. Hays, *Beauty, Health, and Permanence: Environmental Politics in the United States, 1955–1985* (Cambridge: Cambridge University Press, 1987), 545n21.
14. Train, "Environmental Record of the Nixon Administration," 186; Perlstein, *Nixonland*, 460, 544.
15. Clare Conley, "Editorial," *Field & Stream*, April 1972, 4.
16. John Brooks Flippen, "The Nixon Administration, Timber, and the Call of the Wild," *Environmental History Review* 19, no. 2 (1995): 37–41.
17. Perlstein, *Nixonland*, 460; Nelson Rockefeller, chairman, "Commission on CIA

Activities within the United States, Final Report," Washington, DC, June 6, 1975, www.fordlibrarymuseum.gov; "Nixon's Blue-Collar Strategy," *Fredericksburg (VA) Free Lance-Star*, Dec. 27, 1972.
18. Perlstein, *Nixonland*, 497–98.

6. The Sagebrush Rebellion Begins

1. For a discussion of problems *Kleppe* educes, see Karen R. Merrill, *Public Lands and Political Meaning: Ranchers, the Government, and the Property between Them* (Berkeley: University of California Press, 2002), 208–9.
2. Ted Trueblood, "The Forest Service versus the Wilderness Act," *Field & Stream*, Sept. 1975, 16–17, 40, 17; Char Miller, *Public Lands, Public Debates: A Century of Controversy* (Corvallis: Oregon State University Press, 2012), 119–20.
3. William L. Graf, *Wilderness Preservation and the Sagebrush Rebellions* (Savage, MD: Rowman & Littlefield, 1990), 216; Richard Freeman, "The EcoFactory: The United States Forest Service and the Political Construction of Ecosystem Management," *Environmental History* 7, no. 4 (2002): 633–34.
4. Graf, *Wilderness Preservation*, 213–14.
5. Freeman, "EcoFactory," 634.
6. William D. Rowley, *U.S. Forest Service Grazing and Rangelands: A History* (College Station: Texas A&M University Press, 1985), 238.
7. U.S. Department of the Interior, Bureau of Land Management and Office of the Solicitor, eds., Federal Land Policy and Management Act, as amended, 2001, U.S. Department of the Interior, Bureau of Land Management Office of Public Affairs, Washington, DC, sec. 102 (43 U.S.C. 1701)(a), 1; Graf, *Wilderness Preservation*, 221; Robert H. Nelson, "Why the Sagebrush Revolt Burned Out," *Cato Institute* (May–June 1984): 29–30, www.cato.org.
8. R. McGreggor Cawley, *Federal Land, Western Anger: The Sagebrush Rebellion and Environmental Politics* (Lawrence: University Press of Kansas, 1993), 41, 82.
9. James R. Skillen, *The Nation's Largest Landlord: The Bureau of Land Management in the American West* (Lawrence: University Press of Kansas, 2009), 112–13; *San Francisco Examiner & Chronicle*, May 1, 1977.
10. Cecil D. Andrus and Joel Connelly, *Cecil Andrus: Politics Western Style* (Seattle: Sasquatch Books, 1999), 71–72, 78–80.
11. Ibid., 67–69.
12. Cawley, *Federal Land, Western Anger*, 85–86; Stephen Haycox, *Alaska: An American Colony* (Seattle: University of Washington Press, 2002), 295; *Los Angeles Times*, reprinted in "Rebellion or Greed," *Las Vegas Sun*, Sept. 16, 1979.
13. For a short video of the protesters at Cantwell, see www.youtube.com.
14. Graf, *Wilderness Preservation*, 249.
15. Ibid., 225–26; Cawley, *Federal Land, Western Anger*, 92, 110.
16. "Nevada Senators OK 'Sagebrush Rebellion,'" *Sacramento Bee*, May 22, 1979; "First Shots Fired in Sagebrush Rebellion," *Elko (NV) Daily Free Press*, Feb. 16, 1979.
17. Clifton Young, interview by Susan Imswiler, Nov. 30, 1999, "The Sagebrush Rebellion in Nevada," University of Nevada Oral History Program archive, UNOHP Catalog #233, Special Collections, University of Nevada Library, Reno; "Sagebrush Rebel Rally," *Reno Evening Gazette*, April 5, 1979.
18. "Nevada, Rebel without a Plan," *Nevada State Journal*, June 5, 1979.
19. "Nevada Presses Its Effort to Regain Disputed Land," *New York Times*, June 15, 1979.

20. Cawley, *Federal Land, Western Anger*, 109–10; Graf, *Wilderness Preservation*, 230; Joseph M. Chomski, Esq., and Constance E. Brooks, Esq., *Sagebrush Rebellion: A Concise Analysis of the History, the Law and Politics of Public Land in the United States*, prepared for the State of Alaska Legislative Affairs Agency, Legislative Information Office, Anchorage, Jan. 28, 1980, 85, 87, Institute of Governmental Studies Library, University of California, Berkeley.
21. *Los Angeles Times*, reprinted in "Rebellion or Greed."
22. "Nevada Seizes Jurisdiction over US Acres," *Sacramento Bee*, July 15, 1979; Glen D. Weaver, "Nevada's Federal Lands," *Annals of the Association of American Geographers* 59, no. 1 (1969): 35, 41; Graf, *Wilderness Preservation*, 229.
23. *Los Angeles Times*, reprinted in "Rebellion or Greed."
24. "Lawyer Raps Nevada Land Rebellion," *Sacramento Bee*, July 30, 1979.
25. Mike O'Callaghan, "Where I Stand," *Las Vegas Sun*, July 12, 1981; "Secretary Andrus Speaks Out on Sagebrush Rebellion," Nevada BLM news release, Oct. 31, 1979.
26. "Laxalt Answers Andrus on Sagebrush Rebellion," *Elko (NV) Daily Free Press*, Dec. 21, 1979.
27. "Brown Kills Sagebrush Rebellion Bill," *Nevada State Journal*, Sept. 22, 1979.
28. Jedediah Smart Rodgers, *Roads in the Wilderness: Conflict in Canyon Country* (Salt Lake City: University of Utah Press, 2013), 60–61.
29. *Los Angeles Times*, Feb. 16, 1979; *Deseret News*, Nov. 7, 1988.

7. Harnessing Forces

1. "Sagebrush Rebels Falter on Brink of Battle," *Reno Evening Gazette*, July 5, 1980; Joseph M. Chomski, Esq., and Constance E. Brooks, Esq., *Sagebrush Rebellion: A Concise Analysis of the History, the Law and Politics of Public Land in the United States*, prepared for the State of Alaska Legislative Affairs Agency, Legislative Information Office, Anchorage, Jan. 28, 1980, 7, Institute of Governmental Studies Library, University of California, Berkeley.
2. "New Hampshire Congressman Joins Rebellion," *Elko (NV) Independent*, Aug. 6, 1980.
3. Lakes with visibility to a depth of 16 feet are considered clear. In 1873 visibility at Tahoe was measured to 108 feet. Clear-cut logging gravely affected the clarity with silt, sawdust, pine needles, and charcoal from burned slash polluting the lake. Over the course of many decades, Tahoe regained its clarity. In 1968 visibility was measured at 102 feet. With deterioration due to development that began in the '60s, in 1997 it fell to its all-time low, 64.1 feet. With a massive governmental effort to rehabilitate the Tahoe Basin environment, clarity seems to have stabilized. In 2013 the average measurement was 70.1 feet. For information on the history of Lake Tahoe's environmental problems, see Douglas H. Strong, *Tahoe: An Environmental History* (Lincoln: University of Nebraska Press, 1984); and Michael J. Makley, *Saving Lake Tahoe: An Environmental History of a National Treasure* (Reno: University of Nevada Press, 2013).
4. "Tahoe Land Sales Plan Prepared," *Sacramento Bee*, Nov. 10, 1979.
5. John Jacobs, *A Rage for Justice: The Passion and Politics of Phillip Burton* (Berkeley: University of California Press, 1997), 388; Makley, *Saving Lake Tahoe*, 108–9; "Tahoe Land Sales Plan Prepared," *Sacramento Bee*, Nov. 10, 1979.
6. Jacobs, *Rage for Justice*, 393; "Congressional Plans for Lake Tahoe Hit New Snag," *Sacramento Bee*, Jan. 4, 1980.
7. Jacobs, *Rage for Justice*, 392; Makley, *Saving Lake Tahoe*, 113–14.

8. Jedediah Smart Rodgers, *Roads in the Wilderness: Conflict in Canyon Country* (Salt Lake City: University of Utah Press, 2013), 4–6, 64.
9. Eric W. Trenbeath, "The Fight over Lands in Southern Utah Is Shaping Up to Be the Next Sagebrush Rebellion," *Wild Utah*, July 2, 2014; R. McGreggor Cawley, *Federal Land, Western Anger: The Sagebrush Rebellion and Environmental Politics* (Lawrence: University Press of Kansas, 1993), 5.
10. Cawley, *Federal Land, Western Anger*, 5–6.
11. Ibid., 6–7.
12. Henry C. Kenski and Helen M. Ingram, "The Reagan Administration and Environmental Regulation: The Constraint of the Political Market," in *Controversies in Environmental Policy*, edited by Sheldon Kamieniecki, Robert O'Brien, and Michael Clark (Albany: State University of New York Press, 1986), 284–85; William Perry Pendley, *Sagebrush Rebel: Reagan's Battle with Environmental Extremists and Why It Matters Today* (Washington, DC: Regnery, 2013), 10.
13. Rick Perlstein, *The Invisible Bridge: The Fall of Nixon and the Rise of Reagan* (New York: Simon & Schuster, 2014), 393.
14. Paul Laxalt, interview, Miller Center, University of Virginia, Oct. 9, 2001, http://millercenter.org.
15. Ronnie Dugger, *On Reagan: The Man and His Presidency* (New York: McGraw-Hill, 1983), 474–83.
16. Perlstein, *Invisible Bridge*, 455; Kenski and Ingram, "Reagan Administration and Environmental Regulation," 285; Lewis L. Gould, *Grand Old Party: A History of the Republicans* (New York: Random House, 2003), 424. In the mid-1970s Reagan's former associates at General Electric were beginning their multimillion-dollar campaign to fight charges that their PCB emissions had polluted the Hudson River. Between 1947 and 1977 the company had dumped 1.3 million pounds of PCBs into the river, but they contended that the charges were a phony environmentalist controversy invented to destroy jobs. Nationwide, GE is responsible for contamination in seventy-five Superfund sites, the most of any company in the country. "Hudson River PCBs," *Riverkeeper*, www.riverkeeper.org.
17. Gould, *Grand Old Party*, 418; Laxalt, interview.
18. Jerome L. Himmelstein and James A. McRae Jr., "Social Conservatism, New Republicans, and the 1980 Election," *Public Opinion Quarterly* 48, no. 3 (1984): 603–4. The authors used the 1980 Center for Political Studies National Election Study and New York Times / CBS and Gallup Polls.
19. Cawley, *Federal Lands, Western Anger*, 92–93.
20. At the convention Hatch quoted Brigham Young, who admonished his followers to "progress, improve, and make beautiful everything around you." Gary Turbak, "America's Sagebrush Rebellion," *Kiwanis Magazine* (Feb. 1981): 40. William Graf commented, "Preservation of natural areas without conversion of their resources to economic benefits for humans is largely antithetical to the Mormon philosophy. The LDS dominates Utah politics and has a long history of opposing the federal government dating to the 1850s when U.S. troops were sent to the territory which was an LDS theocracy." William L. Graf, *Wilderness Preservation and the Sagebrush Rebellions* (Savage, MD: Rowman & Littlefield, 1990), 238–39.
21. Cawley, *Federal Lands, Western Anger*, 93, 140–41; Ted Stevens, "Speech before the League for Advancement of States' Equal Rights," Nov. 21, 1980, folder 11, box 224, series

Notes to Pages 62–69   137

69, Sierra Club Members Papers, BANC MSS 71/295c, Bancroft Library, Berkeley, CA; Karen R. Merrill, *Public Lands and Political Meaning: Ranchers, the Government, and the Property between Them* (Berkeley: University of California Press, 2002), 205, 234n1; Jacqueline Vaughn Switzer, *Green Backlash: The History and Politics of Environmental Opposition in the U.S.* (Boulder, CO: Lynne Rienner, 1997), 186.

22. "Laxalt on Sagebrush Rebellion," *Nevada State Journal*, Nov. 28, 1980.
23. Cecil D. Andrus and Joel Connelly, *Cecil Andrus: Politics Western Style* (Seattle: Sasquatch Books, 1999), 80–81; Stephen Haycox, *Alaska: An American Colony* (Seattle: University of Washington Press, 2002), 297–99.
24. Andrus and Connelly, *Cecil Andrus*, 226.
25. "Protection for Wild Rivers," *Los Angeles Times*, Jan. 24, 1985. Upon returning to Idaho, Andrus was elected to two more terms as governor, where his legacy includes protecting the Salmon River, the Boulder–White Cloud Area, and the Birds of Prey Area, as well as fighting to restore salmon spawning in the state.

8. Fracturing Policies

1. "Inauguration Address," Jan. 20, 1981, www.presidency.ucsb.edu.
2. William Perry Pendley, *Sagebrush Rebel: Reagan's Battle with Environmental Extremists and Why It Matters Today* (Washington, DC: Regnery, 2013), xxvi, xxviii; Samuel P. Hays, *Beauty, Health, and Permanence: Environmental Politics in the United States, 1955–1985* (Cambridge: Cambridge University Press, 1987), 491–92; James R. Skillen, *The Nation's Largest Landlord: The Bureau of Land Management in the American West* (Lawrence: University Press of Kansas, 2009), 126.
3. Pendley, *Sagebrush Rebel*, 85, 94, 292n69; R. McGreggor Cawley, *Federal Land, Western Anger: The Sagebrush Rebellion and Environmental Politics* (Lawrence: University Press of Kansas, 1993), 113.
4. James Watt confirmation hearing, C-SPAN, Jan. 7, 1981, www.c-span.org.
5. Pendley, *Sagebrush Rebel*, 59–63; "Watt Quits Post," *New York Times*, Oct. 10, 1983.
6. Hays, *Beauty, Health, and Permanence*, 494.
7. Eliot Marshall, "Hit List on EPA?," *Science* 219, no. 4590 (1983): 1303.
8. Paul Sabatier, Susan Hunter, and Susan McLaughlin, "The Devil Shift: Perceptions and Misperceptions of Opponents," *Western Political Quarterly* 40, no. 3 (1987): 450–52, 461, 470; Jacqueline Vaughn Switzer, *Green Backlash: The History and Politics of Environmental Opposition in the U.S.* (Boulder, CO: Lynne Rienner, 1997), 209; James Watt, "The Religious Left's Lies," *Washington Post*, May 21, 2005.
9. Pendley, *Sagebrush Rebel*, xxv–xxviii.
10. Ibid., 4–33, 120–21; Ronnie Dugger, *On Reagan: The Man and His Presidency* (New York: McGraw-Hill, 1983), 86–87.
11. Cawley, *Federal Land, Western Anger*, 114.
12. Skillen, *Nation's Largest Landlord*, 124–25; Cawley, *Federal Land, Western Anger*, 116–17; Switzer, *Green Backlash*, 185.
13. Hays, *Beauty, Health, and Permanence*, 491, 504–5; Cawley, *Federal Land, Western Anger*, 118.
14. Pendley, *Sagebrush Rebel*, 2–5. For an example of how complicated this case became, see *Getty Oil Co. v. Clark*, 614 F. Supp. 904 (D. Wyo. 1985), http://law.justia.com.
15. "Petitions to Oust Watt Go to Congress," *New York Times*, Oct. 20, 1981, Jan. 1, 2001.
16. *Stearns v. Minnesota*, 179 U.S. 223 (1900); *United States v. Texas*, 339 U.S. 707 (1950).

These cases are discussed in Joseph M. Chomski, Esq., and Constance E. Brooks, Esq., *Sagebrush Rebellion: A Concise Analysis of the History, the Law and Politics of Public Land in the United States*, prepared for the State of Alaska Legislative Affairs Agency, Legislative Information Office, Anchorage, Jan. 28, 1980, 110–11, Institute of Governmental Studies Library, University of California, Berkeley.

17. Hays, *Beauty, Health, and Permanence*, 493, 495, 497–98. OMB administrator James Miller said, "There was never any question of the loyalty or effectiveness of the OMB staff. As you radiate out from the White House, there's more of a question. This is especially the case for mission agencies that, in the judgment of the President or his close aides, had gone beyond their mandates, like the social regulators, the EPA, OSHA." James Miller, interview, Miller Center, University of Virginia, Nov. 4, 2001, http://millercenter.org.

18. Miller, interview; "Anne Burford Dies," *Washington Post*, July 22, 2004; Hays, *Beauty, Health, and Permanence*, 496, 515; "Cleaning Up the Chesapeake," *Wilson Quarterly* 11, no. 4 (1987): 65.

19. For a discussion of pressures on agency heads, see former Office of Management and Budget officers Christopher DeMuth and Douglas H. Ginsburg, "Rationalism in Regulation," *Michigan Law Review* 108, no. 6 (2010): 877–912; and Miller, interview.

20. "Anne Burford Dies"; Pendley, *Sagebrush Rebel*, 252n14.

21. "Why James Watt Must Go," *Chicago Tribune*, Sept. 30, 1983; "Watt Tells House Panel He Will Continue Sale of Coal Land Leases," *New York Times*, May 13, 1983; "Watt Battled a Rising Tide," *New York Times*, Oct. 10, 1983.

22. Ronald Reagan, "Message to Congress Transmitting the Fiscal Year 1983 Budget," Feb. 8, 1982, www.reagan.utexas.edu.

23. Steve H. Hanke, "The Privatization Debate: An Insider's View," *Cato Journal* 2, no. 3 (1982): 655–56, 662.

24. Cawley, *Federal Lands, Western Anger*, 118–19; Norm Glaser, interview by Susan Imswiler, "The Sagebrush Rebellion in Nevada," April 9, 1999, University of Nevada Oral History Program archive, UNOHP Catalog #233, Special Collections, University of Nevada Library, Reno.

25. Cawley, *Federal Lands, Western Anger*, 133.

26. Pendley, *Sagebrush Rebel*, 75.

27. Cawley, *Federal Land, Western Anger*, 146–47.

28. Skillen, *Nation's Largest Landlord*, 127–28; Joseph Ross, "FLPMA Turns 30: The Bureau of Land Management Also Celebrates Its 60th birthday," *Society for Range Management* (Oct. 2006): 21, www.blm.gov/flpma/; Dean Rhoads, interview by Imswiler, "Sagebrush Rebellion in Nevada," Sept. 4, 1999, University of Nevada Oral History Program archive; Hays, *Beauty, Health, and Permanence*, 499–500.

29. Hays, *Beauty, Health, and Permanence*, 502.

30. Richard Freeman, "The EcoFactory: The United States Forest Service and the Political Construction of Ecosystem Management," *Environmental History* 7, no. 4 (2002): 634–36.

31. Hal K. Rothman, *Saving the Planet: The American Response to the Environment in the Twentieth Century* (Chicago: Ivan R. Dee, 2000), 170.

32. James Morton Turner, "The Specter of Environmentalism: Wilderness, Environmental Politics, and the Evolution of the New Right," *Journal of American History* 96, no. 1 (2009): 123, 125, 136; "1984 Scorecard Vote, Burford Appointment," League of Conservation Voters, http://scorecard.lcv.org.

9. Inciting the Populace

1. Samuel P. Hays, *Beauty, Health, and Permanence: Environmental Politics in the United States, 1955–1985* (Cambridge: Cambridge University Press, 1987), 506; Samuel P. Hays, "Environmental Political Culture and Environmental Political Development: An Analysis of Legislative Voting, 1971–1989," *Environmental History Review* 16, no. 2 (1992): 3–6; Paul Larmer, "Congress' War against Nature Creates Backlash," *High Country News*, Dec. 11, 1995.
2. Edward Abbey, "Cowburnt," *Harper's Magazine*, Feb. 2015, 33, reprinted from *Harper's*, Jan. 1986.
3. R. McGreggor Cawley, *Federal Land, Western Anger: The Sagebrush Rebellion and Environmental Politics* (Lawrence: University Press of Kansas, 1993), 156–59; Richard Freeman, "The EcoFactory: The United States Forest Service and the Political Construction of Ecosystem Management," *Environmental History* 7, no. 4 (2002): 636.
4. Richard White, "The Current Weirdness in the West," *Western History Quarterly* 28, no. 1 (1997): 12.
5. James Morton Turner, "The Specter of Environmentalism: Wilderness, Environmental Politics, and the Evolution of the New Right," *Journal of American History* 96, no. 1 (2009): 132, 137–41.
6. Ralph Maughan and Douglas Nilson, "What's Old and What's New about the Wise Use Movement," paper delivered at Western Social Science Association Convention, Corpus Christi, TX, April 23, 1993, http://aalto.arch.ksu.edu; Jacqueline Vaughn Switzer, *Green Backlash: The History and Politics of Environmental Opposition in the U.S.* (Boulder, CO: Lynne Rienner, 1997), 210–11; William Kevin Burke, "The Wise Use Movement: Right-Wing Anti-Environmentalism," *Public Eye* 7, no. 2 (1993), www.publiceye.org.
7. Hal K. Rothman, *Saving the Planet: The American Response to the Environment in the Twentieth Century* (Chicago: Ivan R. Dee, 2000), 177–79; Switzer, *Green Backlash*, 197; David Helvarg, "Wise Use in the White House: Yesterday's Fringe, Today's Cabinet Official," *Sierra Magazine*, Sept.–Oct. 2004, http://vault.sierraclub.org.
8. "War of Words," *Los Angeles Times*, Jan. 21, 1991; Dan Nimmo and Chevelle Newsome, *Political Commentators in the United States in the Twentieth Century: A Bio-critical Sourcebook* (Westport, CT: Greenwood Press, 1997), 186.
9. Freeman, "EcoFactory," 643–46; Republican Party Platform of 1992, Aug. 17, 1992, American Presidency Project, www.presidency.ucsb.edu.
10. White, "Current Weirdness in the West," 8; Switzer, *Green Backlash*, 209–10, 238; Timothy Egan, "Terror in Oklahoma: In Congress; Trying to Explain Contacts with Paramilitary Groups," *New York Times*, May 2, 1995. Chenoweth had the ability to inspire and mobilize supporters, but she was a particularly ineffective lawmaker. In her three terms as a representative, she introduced forty-four bills—none of which made it out of committee.
11. Helvarg, "Wise Use in the White House"; William Chaloupka, "The County Supremacy and Militia Movements: Federalism as an Issue on the Radical Right," *Publis* 26, no. 3 (1996): 169; Switzer, *Green Backlash*, 200–205, 215–16; Rothman, *Saving the Planet*, 179–80.
12. Chaloupka, "County Supremacy and Militia Movements," 162–63. Catron County is part of the nineteenth- and early-twentieth-century empire of Thomas B. Catron. By 1880 his holdings were the largest in the history of the United States, more than three million acres. Patricia Nelson Limerick, *The Legacy of Conquest: The Unbroken Past of the American West* (New York: W. W. Norton, 1987), 237–38.

13. Rothman, *Saving the Planet*, 180–81; Timothy Egan, "Court Puts Down Rebellion over Federal Land," *New York Times*, March 16, 1996; Frankie Sue Del Papa, Nevada Attorney General, Carson City, to All State Legislators, District Attorneys, and County Commissioners, March 3, 1994. Asked about his constituent groups' apparent racism "and vitriolic antigovernment rhetoric," Nye County's Carver said he disavowed such language, "but feels that as an elected official, he cannot discriminate against any audience just because its views are more extreme than his" (Chaloupka, "County Supremacy and Militia Movements," 170).
14. David Helvarg, *The War against the Greens: The Wise Use Movement, the New Right and the Browning of America* (Boulder, CO: Johnson Books, 2004), 276–77; Switzer, *Green Backlash*, 230.
15. Switzer, *Green Backlash*, 219.
16. Helvarg, *War against the Greens*, 275; Switzer, *Green Backlash*, 216–17.
17. Switzer, *Green Backlash*, 217; Helvarg, *War against the Greens*, 276–77; Chaloupka, "County Supremacy and Militia Movements," 170.
18. "The Emotional Politics of a Political Trial," *New York Times*, April 27, 1995.
19. Antonio de Velasco, *Centerist Rhetoric: The Production of Political Transcendence in the Clinton Presidency* (Lanham, MD: Rowman & Littlefield, 2010), 102.
20. "Speaker Speaketh Too Much," *Los Angeles Times*, May 9, 1995; Chaloupka, "County Supremacy and Militia Movements," 170.
21. Chaloupka, "County Supremacy and Militia Movements," 168; "U.S. Studies Wave of Violence in Nevada," *New York Times*, Dec. 22, 1995.

10. County versus Federal Government

1. James W. Hulse, *The Nevada Adventure: A History*, 6th ed. (1965; reprint, Reno: University of Nevada Press, 1990), 2–9; Leisl Carr Childers, *The Size of the Risk: Histories of Multiple Use in the Great Basin* (Norman: University of Oklahoma Press, 2015), 22; Dean Rhoads, interview by Susan Imswiler, "The Sagebrush Rebellion in Nevada," Sept. 4, 1999, University of Nevada Oral History Program archive, UNOHP Catalog #233, Special Collections, University of Nevada Library, Reno.
2. Florence Williams, "The Shovel Brigade," *Mother Jones*, Jan.–Feb. 2001, www.motherjones.com; Childers, *Size of the Risk*, 147.
3. Williams, "The Shovel Brigade"; *Idaho Statesman* (Boise), Nov. 3, 2014.
4. "John Ensign Backs Jarbidge Shovel Brigade," *Las Vegas Sun*, Feb. 16, 2000.
5. Williams, "The Shovel Brigade."
6. Marjorie Sill, interview by Imswiler, "Sagebrush Rebellion in Nevada," Oct. 26, 1999, University of Nevada Oral History Program archive; Bill Kohlmoos, interview by Imswiler, "Sagebrush Rebellion in Nevada," June 22, 2001, ibid.
7. Gloria Flora, interview by Imswiler, "Sagebrush Rebellion in Nevada," Sept. 15, 2000, ibid.
8. The Great Old Broads for Wilderness had organized in 1989 after taking umbrage over an Orrin Hatch quote. He said the elderly needed more roads and paved ways to get into wilderness, and they disagreed. Susan Tixier, interview by Imswiler, Sept. 17, 2000, "Sagebrush Rebellion in Nevada," ibid.; Williams, "The Shovel Brigade."
9. "Liberty Rock Yields to Shovelers," *Los Angeles Times*, July 5, 2000; Tixier, interview.
10. "Shovel Brigade Explains Basis for Comments," *Elko (NV) Daily Free Press*, Oct. 24, 2003.

11. "Judge Refuses to Toss 'Shovel Brigade' Deal," *Elko (NV) Daily Free Press*, Aug. 20, 2012; Williams, "The Shovel Brigade"; Nathan Brown, "15 Years after Shovel Brigade, Debate on Federal Control Rages On," May 31, 2015, http://magicvalley.com.

11. Extinguished Rights

1. Scott Robert Ladd, "Stealing Nevada," Coyote Gulch Productions, n.d., www.dickshovel.com; Rebecca Solnit, *Savage Dreams: A Journey into the Hidden Wars of the American West* (Berkeley: University of California Press, 2014), 194–96.
2. *Mary and Carrie Dann v. United States*, Case 11.140, Report No. 75/02, Inter-Am. C.H.R., Doc. 5 rev. 1 at 860 (2002), University of Minnesota, Human Rights Library, www1.umn.edu/humanrts; "Indian Land Feud Heats Up," *Los Angeles Times*, Sept. 24, 1991; "A Fair Settlement, or Just Settling?" *Washington Post*, Aug. 26, 2002; "U.S. Agents Seize Horses of Defiant Indian Sisters," *New York Times*, Feb. 7, 2003.
3. Solnit, *Savage Dreams*, 159–60.
4. *Mary and Carrie Dann v. United States*; Stephen J. Crum, *The Road on Which We Came: A History of the Western Shoshone* (Salt Lake City: University of Utah Press, 1994), 24–26; Deborah Schaaf and Julie Fishel, "*Mary and Carrie Dann v. United States* at the Inter-American Commission on Human Rights: Victory for Indian Land Rights and the Environment," *Tulane Environmental Law Journal* 16, no. 1 (2002): 179.
5. Crum, *Road on Which We Came*, 24; Ned Blackhawk, *Violence over the Land: Indians and Empires in the Early American West* (Cambridge, MA: Harvard University Press, 2006), 267–68.
6. Schaaf and Fishel, "*Mary and Carrie Dann*," 180–81; Solnit, *Savage Dreams*, 186; "Fair Settlement, or Just Settling?"; Allison M. Dussias, "Squaw Drudges, Farm Wives, and Dann Sisters' Last Stand: American Indian Women's Resistance to Domestication and the Denial of Their Property Rights," *North Carolina Law Review* (Jan. 1999): 724.
7. *Western Shoshone v. United States of America*, before the Indian Claims Commission, Docket No. 326-K, Oct. 11, 1972, http://digital.library.okstate.edu.
8. Schaaf and Fishel, "*Mary and Carrie Dann*," 180–81; Crum, *Road on Which We Came*, 176–79, 181.
9. *Western Shoshone Tribal Council et al. v. United States*, U.S. Court of Appeals for the Federal Circuit, U.S. Court of Federal Claims, 2007-5020 (2008), https://turtletalk.files.wordpress.com.
10. Ladd, "Stealing Nevada"; *Los Angeles Times*, Sept. 24, 1991.
11. Solnit, *Savage Dreams*, 179.
12. *Los Angeles Times*, Sept. 24, 1991.
13. Solnit, *Savage Dreams*, 217–18; Crum, *Road on Which We Came*, 182–83.
14. *Mary and Carrie Dann v. United States*; Blackhawk, *Violence over the Land*, 287–88.
15. *Washington Post*, Aug. 26, 2002; *Mary and Carrie Dann v. United States*.
16. *Elko (NV) Daily Free Press*, March 4, 2011.
17. Stacy L. Leeds and Elizabeth Mashie Gunsaulis, "Gender, Justice, and Indian Sovereignty: Native American Women and the Law," *Thomas Jefferson Law Review* (Spring 2012): 303, 321–22; *Elko (NV) Daily Free Press*, April 24, 2005.

12. Pursuing Ideology in the Courts

1. *The Estate of E. Wayne Hage, et al v. United States*, Supreme Court, No. 12-918 (May 2013); Leisl Carr Childers, "The Angry West: Understanding the Sagebrush Rebellion

in Rural Nevada," in *Bridging the Distance: Common Issues of the Rural West*, edited by David B. Danbom (Salt Lake City: University of Utah Press, 2015), 230. For a perception of the difference between Hage's perspective and that of a longtime Nevada ranching family, see Childers, "Angry West," 230–33.
2. *The Estate of E. Wayne Hage, et al v. United States*; Larry Schmidt, taped phone interview by the author, Feb. 10, 2016, tape in possession of the author.
3. Wayne Hage, *Storm over Rangelands: Private Rights in Federal Lands* (Bellevue, WA: Free Enterprise Press, 1989), 207, 223–24.
4. *The Estate of E. Wayne Hage, et al. v. United States.*
5. Wayne Hage, interview by Susan Imswiler, "The Sagebrush Rebellion in Nevada," Sept. 3, 2001, University of Nevada Oral History Program archive, UNOHP Catalog #233, Special Collections, University of Nevada Library, Reno.
6. Hage, *Storm over Rangelands*, 237.
7. Ibid., 211–12, 214.
8. Hage, interview; Hage, *Storm over Rangelands*, 67–68, 192–94, 211.
9. Jacqueline Vaughn Switzer, *Green Backlash: The History and Politics of Environmental Opposition in the U.S.* (Boulder, CO: Lynne Rienner, 1997), 251, 266.
10. *The Estate of E. Wayne Hage and the Estate of Jean N. Hage v. United States*, U.S. Court of Appeals for the Federal Circuit, case no. 91-CV-1470, July 26, 2012.
11. *The Estate of E. Wayne Hage v. United States: Petition for Certiorari*, denied on June 17, 2013, www.scotusblog.com.
12. *United States of America v. Estate of E. Wayne Hage*, U.S. Court of Appeals for the Ninth Circuit, No. 13-16974 D.C. No. 2:07-cv-01154-RCJ-VCF Opinion.
13. Ibid.
14. "Nevada Ranching Family Loses Federal Lands Court Case," *Nevada Review-Journal*, Jan. 18, 2016.

13. Pursuing Ideology with Guns

1. Reuters, "Before Cliven Bundy Stand-Off, a Collision between Ranchers and Tortoises," May 30, 2014; "BLM Releases Bundy Cattle after Protestors Block State Highway I-15," *Las Vegas Review-Journal*, April 12, 2014.
2. Reuters, "Nevada Cattle Rancher Calls on Local Sheriffs to Join His Cause," April 14, 2014.
3. "Cliven Bundy Saga Forces Republicans into Awkward U-Turn from Far Right," *Guardian*, April 25, 2014.
4. "Bundy Ranch Standoff Was Highly Coordinated, Reflecting Threat of Larger Far-Right Militia Movement," Southern Poverty Law Center Report, July 10, 2014, www.splcenter.org; Christopher Ketcham, "The Great Republican Land Heist: Cliven Bundy and the Politicians Who Are Plundering the West," *Harper's Magazine*, Feb. 2015, 26; "Recapture Canyon, Utah, to Be Next Site of BLM Showdown," *Denver Post*, May 3, 2014.
5. Ketcham, "Great Republican Land Heist," 29.
6. Associated Press, "BLM Closes Office as Armed Protesters Gather Outside in Gold Mine Dispute," April 24, 2015.
7. "Commissioner Lyman, Monte Wells Convicted for Role in Capture Canyon Protest," *San Juan Record*, May 1, 2015; "Judge Sentences San Juan County Commissioner to 10 Days in Jail, 3 Years of Probation," *Deseret News*, Dec. 18, 2015.
8. "Who Wants a Burns Standoff?," Jan. 1, 2016; "Gunmen Hold Federal Building in Ore-

gon," Jan. 4, 2016; and "Refuge Militants to Be Arraigned," Feb. 23, 2016, all Oregon Public Broadcasting, www.opb.org.
9. "Protesters Vow to Hold Wildlife Office Indefinitely," *Idaho Statesman* (Boise), Jan. 4, 2016; "Bundy Says He's on a Mission from God," *Idaho Statesman*, Jan. 6, 2016; "Rancher Killed as FBI Arrests Bundys, Others," *Idaho Statesman*, Jan. 27, 2016; "Gunmen Hold Federal Building"; "Malheur Refuge Manager: It's 1 Big Mess," Oregon Public Broadcasting, March 1, 2016, www.opb.org.
10. "Standoff in Oregon Attracts Supporters Bearing Disparate Grievances," *New York Times*, Jan. 17, 2016.
11. Dana Milbank, "Sympathy for Sedition?," *Washington Post*, Jan. 7, 2016.
12. Rocky Barker, "Takeover Goal: A New Government," *Idaho Statesman* (Boise), Jan. 26, 2016; "Rancher Killed as FBI Arrests Bundys, Others"; *United States v. Ammon Bundy, Jon Ritzheimer et al.*, U.S. District Court District of Oregon, Criminal Complaint Case 3:16-my-00004-1,2,3,4,5,6,7,8 (Jan. 26, 2016).
13. "Arrest Adds to Tensions," *Idaho Statesman* (Boise), Jan. 21, 2016.
14. *New York Times*, Jan. 7, 2016; Oregon Public Broadcasting, March 2, 2016, www.opb.org.
15. "4,000 Artifacts Stored at U.S. Wildlife Refuge Held by Armed Group," *Idaho Statesman* (Boise), Jan. 21, 2016.
16. Michele Fiore, Assemblywoman District 4, http://votefiore.com; Oregon Public Broadcasting, Feb. 8, 11, 2016, www.opb.org.
17. "House Approves Measure to Shield Officer Who Shot LaVoy Finicum," Oregon Public Broadcasting, Feb. 17, 2016, www.opb.org.

14. Opaque Governance

1. James R. Skillen, *The Nation's Largest Landlord: The Bureau of Land Management in the American West* (Lawrence: University Press of Kansas, 2009), 165, 181, 259n2.
2. Karen Dodwell, "The Cowboy Myth, George W. Bush, and the War with Iraq," March 2004, www.americanpopularculture.com.
3. Skillen, *Nation's Largest Landlord*, 184.
4. "Document Says Oil Chiefs Met with Cheney Task Force," *Washington Post*, Nov. 16, 2005; "Papers Detail Industries Role in Cheney's Energy Report," *Washington Post*, July 18, 2007; Scott Horton, "Six Questions for Bart Gellman, Author of *Angler*," *Harper's*, Sept. 17, 2008, http://harpers.org/blog. For a discussion of Cheney's management style as vice president, see Barton Gellman and Jo Becker, "Angler: The Cheney Vice Presidency—'a Different Understanding with the President,'" *Washington Post*, June 24–27, 2007.
5. Fred Powledge, "Environmental Science after Bush," *Bio Science* 59, no. 3 (2009): 200–204.
6. "Salmon Experts Pressured to Change Findings," Union of Concerned Scientists, n.d., www.ucsusa.org.
7. Ibid.; Powledge, "Environmental Science after Bush."
8. Sheldon Rampton, "Fish Out of Water: Behind the Wise Use Movement's Victory in Klamath," *PR Watch* 11, no. 1 (2004).
9. "U.S. Denies Blame for Die-Off of Salmon," *Los Angeles Times*, Oct. 3, 2002.
10. "As Thousands of Salmon Die, Fight for River Erupts Again," *New York Times*, Nov. 28, 2002.
11. Ibid.

12. Sharon Levy, "Turbulence in the Klamath River Basin," *BioScience* 53, no. 4 (2003): 315–20.
13. "Salmon Experts Pressured to Change Findings."
14. Frank Luntz, "Straight Talk" memorandum to Bush White House, 2002, www2.bc.edu.
15. "Faith, Certainty and the Presidency of George W. Bush," *New York Times Magazine*, Oct. 17, 2004; "Christie Whitman's Troubled Tenure," *New York Times*, May 22, 2003; Powledge, "Environmental Science after Bush"; "Bush's EPA Is Pursuing Fewer Polluters," *Washington Post*, Sept. 30, 2007.
16. Skillen, *Nation's Largest Landlord*, 166, 184–85; Powledge, "Environmental Science after Bush."
17. In 2006 Jack Abramoff pleaded guilty to federal charges of fraud and tax evasion after he defrauded four Native American tribes out of more than $23 million while advising them regarding operating Indian casinos. Abramoff, who had involved various public officials in his endeavors, was sentenced to forty-eight months in prison and three years of supervised release and was ordered to pay $23,134,695 in restitution. See "Former Lobbyist Jack Abramoff Sentenced to 48 Months in Prison on Charges Involving Corruption, Fraud, Conspiracy and Tax Evasion," Sept. 4, 2008, www.justice.gov.
18. Skillen, *Nation's Largest Landlord*, 167–71.
19. "In Brief: Where They Stand on Science Policy," *Science* 322, no. 5901 (2008): 518.

15. Differing Values

1. Daniel Schulman, *Sons of Wichita: How the Koch Brothers Became America's Most Powerful and Private Dynasty* (New York: Grand Central, 2014), 5–6, 308–10.
2. Seth Shulman, editorial director, Union of Concerned Scientists, "University of Kansas Case Exposes Koch Campus Strategy," *Huffington Post*, Sept. 22, 2015, www.huffingtonpost.com; Jane Mayer, "Covert Operations," *New Yorker*, Aug. 30, 2010, www.newyorker.com; Jane Mayer, *Dark Money: The Hidden History of the Billionaires behind the Rise of the Radical Right* (New York: Doubleday, 2016), 153–54.
3. Schulman, *Sons of Wichita*, 219–21, 249.
4. Polluter Watch, http://polluterwatch.com/koch-industries.
5. Schulman, *Sons of Wichita*, 265, 280, 296; Al Gore, *The Assault on Reason* (New York: Penguin Press, 2007), 201.
6. "Charles Koch Invokes Fight for Civil Rights as Model for Political Activism," *Washington Post*, Aug. 2, 2015.
7. Mayer, *Dark Money*, 346; Brian Murphy, "Meet ALEC's (Hoped for) Man in Washington: Scott Walker," *Talking Points Memo*, Aug. 10, 2015, http://talkingpointsmemo.com.
8. Mayer, *Dark Money*, 370.
9. Christopher Ketcham, "The Great Republican Land Heist: Cliven Bundy and the Politicians Who Are Plundering the West," *Harper's Magazine*, Feb. 2015, 27–28.
10. S.Amdt. 838 to S.Con.Res.11, 114th Cong., March 25, 2015, www.congress.gov.
11. "Idaho Freedom Foundation's Charitable Status Scrutinized," *Spokesman Review* (Spokane, WA), Sept. 15, 2013.
12. Holly Fretwell and Shawn Regan, "Divided Lands: State vs. Federal Management in the West," Property and Environment Research Center, 2015, www.perc.org; *Idaho Statesman* (Boise), March 22, 2013, April 17, 2014.
13. "The Wildfire Burden: Why Public Land Seizure Proposals Would Cost Western States Billions of Dollars," http://westernpriorities.org; Evan Hjerpe, "Fiscal Impacts to the

State of Idaho from HR 22 Implementation," Idaho Conservation League, Dec. 2013, www.idahoconservation.org.
14. Fretwell and Regan, "Divided Lands," 12, 19.
15. Fred Birnbaum, "Criticism of Transfer Idea Is Sign of Traction," *Idaho Statesman* (Boise), April 16, 2015.
16. Roger Phillips, "Our Lands Deserve Protection, Not Liquidation," *Idaho Statesman* (Boise), April 16, 2015.
17. Jonathan Oppenheimer, "Legislature Should Stop Futile Exercise," *Idaho Statesman* (Boise), Feb. 3, 2015.

Conclusion

1. Chance Finegan, "The Value of Public Lands: A Broad, Quiet Consensus," *Public Lands / Blog West,* April 20, 2015, https://blogwest.org.
2. "Crapo's Vote on Transfer Upsets Collaborators," *Idaho Statesman* (Boise), April 8, 2015.
3. "Republicans Slam Obama's Latest 'Land Grab,'" *Washington Times,* July 10, 2015; "Our Environment," White House, www.whitehouse.gov.
4. Matt Lee-Ashley, "Ted Cruz Vows to Sell Off or Give Away Nevada's Public Lands," Think Progress, Feb. 19, 2016, http://thinkprogress.org; Act of Congress (1864) Enabling the People of Nevada to Form a Constitution and State Government, www.leg.state.nv.us.
5. "Donald Trump Blasts Environmental Protection Agency," *Bloomberg Politics,* Oct. 18, 2015, www.bloomberg.com.

Index

Abbey, Edward, 54, 74–75
Abramoff, Jack, 114, 144n17
Acheson, Dean, 32
Agnew, Spiro, 39
Alaska, 4, 32; commerce disputes in, 18–20, 41, 49–50, 62–63; Native claims in, 132n27; Sagebrush Rebellion and, 52, 54; wilderness in, 49–50, 62–63
Alaska Legislative Affairs Agency, 55
AMAX (mining), 70
American Enterprise Institute, 50
American Legislative Exchange Council (ALEC), 117–18
American River, 63
Anderson, Clinton P., 29
Andrus, Cecil, 48–49, 53, 62
Arctic National Refuge, 113
Arizona, 4, 32, 122; extremists and, 81; Goldwater and, 31–32, 35; revenue in, 7, 32, 119–20, 131n8; Sagebrush Rebellion and, 52
Arnold, Ron, 76–77, 79–81, 95
Ashcroft, John, 116
Aspinall, Wayne, 35, 37, 132n22
Atlantic Monthly, 29

Babbitt, Bruce, 52, 108
Baden, John, 62
Bagge, Carl, 48
Ballinger, Richard, 16–20
Barrett, Frank, 27–29

Billings, Montana, 70–71
Black, Calvin, 54
Block, Herbert L., 64
Bob Marshall Wilderness area, 68
Boulware, Lemuel, 59
Brandborg, Stewart M., 36
Brimmer, Clarence, 108
Brock, J. Elmer, 29, 130n21
Brower, David, 30–31, 65
Brown, Jerry, 53–54
Bull Moose Party, 20
Bundy, Ammon, 3, 100, 103–5, 107
Bundy, Cliven, 2–4, 101–2, 107
Bundy, Ryan, 100, 102, 106–7
Bureau of Land Management (BLM), 33; Bundys and, 2–3, 101–3, 105, 107; Danns and, 88–89, 92–94; firefighting and, 7; formation of, 27, 33; funding of, 27, 33, 120; George W. Bush administration and, 113–14; Hage and 99; MUSY and, 32; Nevada and, 51–52, 56–57, 83, 87; Reagan administration and, 66–67, 72; roads and, 58; Sagebrush Rebellion and, 51–52; Sierra Club and, 31; Utah and, 54, 58, 102
Bureau of Indian Affairs, 90, 92, 93
Bureau of Reclamation, 17, 31
Burford, Robert, 66, 72
Burns, Oregon, 105–7
Burns Paiute Tribe, 105

Burton, Phillip, 56–57
Bush, George H. W., 77, 108
Bush, George W., 108–9, 112–14
Butz, Earl, 42, 46

California, 4; interior department and, 18; Lake Tahoe and, 18, 56; metropolitan shift in, 32; Muir and, 30; Reagan and, 39; Sagebrush Rebellion and, 52; rivers and, 63; timber and, 67
California Chamber of Commerce, 50
California Legislature, 18
Cannon, "Uncle Joe," 11, 14
Cantwell, Alaska, 50
Carhart, Arthur, 28
Carpenter, Farrington, 23–24, 25, 131n9
Carson, Rachel, 36
Carter, Jimmy, 40, 48–49, 50, 57, 61, 63–64, 69
Carver, Richard, 79, 140n13
Catron County, New Mexico, 78–79, 80, 144n12
Center for Defense of Free Enterprise, 80
Chaffetz, Jason, 118
Cheney, Dick, 68, 109–10, 113, 115
Chenoweth, Helen, 77, 144n10
Civil War, 10, 90
Clark, William P., 72
Clark County, Nevada, 57, 101–2
Clarke, Kathleen, 114
Clawson, Marion, 62
Clean Air Act, 40, 60, 113, 117
Cleveland, Grover, 10
Clinton, Bill, 74, 77, 81, 108, 109
Coldiron, William, 70–71
College of Colorado, 122
Collier's (magazine), 29
Colorado, 4, 14, 52, 53, 122
Colorado River Storage Project, 31
Constitution of the United States, 9, 45
Coolidge, Calvin, 20
Coors, Joseph, 50, 64
Council of Economic Advisers (Reagan's), 7, 71
County Supremacists, 78–79, 100–101, 104

County Supremacy movement, 75, 78–79. *See also* County Supremacists
Court of Federal Claims (CFC), 97–98
Cripple Creek, 13
Crowell, John, 66, 72
Cruz, Ted, 3, 7, 123
Cunningham, Clarence, 18–19

Dann, Carrie, 88–90, 92–93
Dann, Clifford, 88–89
Dann, Mary, 90, 93
Debs, Eugene, 20
Del Papa, Frankie Sue, 79
Democratic Party, 5–6, 35, 43, 49, 60–62, 74, 77, 118
Denver Post, 29
Dern, George H., 21
Deseret News, 54
Devaney, Earl, 114
DeVoto, Bernard, 28, 53
Dionne, E. J., 81
"Don't Tread on Me," (Gadsden flag), 50, 102
Dudley, Susan, 116

Earth First!, 72, 75, 79–80
East Meadow Creek, Colorado, 46–47
Ecodefense: A Field Guide to Monkeywrenching (Foreman), 75
Eel River, 63
Eisenhower, Dwight D., 31–32
Elko County, Nevada, 83–87
Elko Daily Free Press, 51, 86
Endangered Species Act, 44, 47, 50, 77
Ensign, John, 84
Environmental Protection Agency (EPA): ALEC attacks on, 118; extremists and, 103; George H. W. Bush and, 77; George W. Bush and, 108, 113; Mercatus Institute's attacks on, 115–116; Nixon creation of, 44; Reagan administration and, 60, 66, 69–70, 72; Trump on, 123
Evers, Medgar, 38
ExxonMobile, 117

Fall, Albert B., 20
Federal Bureau of Investigation (FBI), 60, 106–7
Federal Land Policy and Management Act (FLPMA), 47–48
"Federal Real Property Initiative," 71
Field & Stream, 46
Fifth Amendment, 91, 94, 97
Finecum, Robert "LaVoy," 106
Fiore, Michele, 107
Fisher, Walter L., 19–20
Flora, Gloria, 85
Ford, Gerald, 60
Foreman, Dave, 75
Forest Reserve Act of 1891, 10
Fulton, Charles, 14

Garfield, James, Jr., 16–17, 21
General Electric, 59, 139n16
George Mason University, 115
Gila Wilderness, 34
Gingrich, Newt, 77, 81
Glaser, Norm, 51
Glavis, Louis R., 18–19
Golden Gate National Recreation Area, 57
Goldwater, Barry, 31–32, 35, 39, 50, 131n8
Gorsuch, Anne, 66, 69–70
Grand Canyon, 17, 31, 33, 36, 49
Grand County, Utah, 54
Grants Pass, Oregon, 103
Grasty, Steve, 105
Gravel, Mike, 49
Great Basin, 83, 89, 119
Great Denali Trespass, 50
Great Old Broads for Wilderness, 86–87, 146n8
Great Salt Lake, 89
Grijalva, Raúl, 118
Griles, J. Steven, 114

Hage, Wayne, 95–99, 141–42n1
Hage, Wayne, Jr., 96, 100
Hamilton, Alexander, 9
Hanke, Steve H., 7, 71
Hankins, Helen, 93
Hannity, Sean, 3

Harding, Warren G., 20
Harmer, John L., 50, 59
Harney County, Oregon, 103–5
Harris, Fred, 42
Harris, Robert, 66
Harrison, Benjamin, 10
Hatch, Orrin, 50, 52, 54–55, 62, 68–69, 139n20
Hawaii, 4, 32, 52
Hays, Samuel P., 72
Heard, Dwight B., 14
Heller, Dean, 3
Henry, Patrick, 9
Heritage Foundation, 64, 66
Hickel, Walter, 37, 41, 43
Hogg, Hershel, 13
Holford, Matt, 86
Hoover, Herbert, 21
Hoover Commission on Public Lands, 21
Humboldt-Toiyabe National Forest, 85
Hunter, Susan, 66
Hunter v. United States, 97, 99

Ickes, Harold L., 22–26, 132n2, 131n9
Indian Claims Commission (ICC), 90–91, 93
Inter-American Commission on Human Rights, 93

Jackson, Andrew, 9
Jackson, Henry "Scoop," 49
Jackson, Leroy, 78
Jackson, Mississippi, 43
Jackson Hole, Wyoming, 68
Jackson State College, 43
Jarbidge, Nevada, 84–87
Jastro, H. A., 14
Jewell, Sally, 104
Johnson, Jed, 27
Johnson, Lyndon, 35–37, 39
Johnson, Stephen L., 113
Jones, Robert Clive, 99
J. P. Morgan–Guggenheim Company, 19

Kaiser, Ruth, 79
Karuk Tribe, 111
Kelly, Mike, 111–12

Kennedy, John F., 5, 32–33, 35–36
Kent, William, 18
Kent State University, 43
Keppen, Dan, 112
Kerr, Andy, 78
King, Martin Luther, 39
Klamath Bucket Brigade, 110
Klamath Irrigation Project, 110–12
Klamath River, 63, 110–12
Klamath Water Users Association, 112
Kleppe v. New Mexico, 45
Koch, Charles, 115–18
Koch, David, 115–16
Koch Industries, 115–18
Kohlmoos, Bill, 85

Labrador, Raul, 105
"land ethic," 34, 38
Lara, Walt, 111
Larson, Arthur, 32
Las Vegas, 56–57, 101
Laxalt, Paul, 53, 56, 59, 61, 62
League for the Advancement of States' Equal Rights (LASER), 50, 59, 61–62, 68
League of Conservation Voters, 6, 42, 108
League of Women Voters, 36, 42
Lecky, Jim, 111–12
Leopold, Aldo, 34, 38
Liddy, G. Gordon, 81
Limbaugh, Rush, 77, 81
Lincoln, Abraham, 10
Los Angeles Herald, 17, 19
Los Angeles Times, 53, 81
Lujan, Manual, Jr., 92
Luntz, Frank, 112–13
Lyman, Phil, 102–3

MacDonald, Julie A., 114
Mack, Richard, 2
Madison, James, 9
Malheur National Wildlife Refuge, 105–6
Marble Canyon dam, 33
Mason, George, 9
Massie, Tom, 104

Mastern, Sue, 112
McCarran, Pat, 25–28, 133n22, 133n1
McGerr, Michael, 31
McKibben, Howard, 89
McLaughlin, Susan, 66
Mercatus Institute, 115
Militia of Montana, 80
Miller, Terry, 52
Moab, Utah, 54, 57–58, 59
Mojave Desert, 80, 101
Monkey Wrench Gang, The (Abbey), 54, 75
Montana, 4, 7, 14, 46, 52, 119–20, 122
Montana Power Company, 70
Mormon Church, 104, 139n20
Morton, Rogers, 41, 43, 47
Mountain States Legal Defense Fund, 50, 65, 78, 95, 98
Muir, John, 12, 30, 38
Multiple Use and Sustained Yield (MUSY) Act, 32
Murkowski, Lisa, 118, 119, 122

National Advisory Board Council, 26
National Cattlemen's Beef Association, 110
National Environmental Policy Act (NEPA), 40, 41
National Federal Lands Conference (NFLC), 79
National Legal Center for the Public Interest, 50
National Marine Fisheries Service (NMFS), 110–12
National Mining Association, 109
National Parks and Conservation Association, 80
National Park Service, 42, 50, 76
National Research Council (NRC), 110–12
National Scenic Trails Act, 35
National Wool Growers Association, 27
Natural Resources Defense Council v. Morton, 47
Nevada, 4, 32, 83; court decisions and, 24, 99–100; Elko County and, 84–85; extremists and, 82, 107; grazing in,

25, 130n1; Las Vegas and, 56; Mojave Desert tortoise and, 101; Nye County and, 79; public lands in, 6, 53, 56–57, 83, 88, 90, 95, 123; Sagebrush Rebellion in, 51–53, 55; Shoshone land and, 88, 90
Nevada Cattlemen's Association, 96
Nevada Department of Environmental Protection, 84
Nevada Department of Wildlife, 95
Nevada State Journal, 52
New Deal, 21, 23, 32
New Mexico, 4, 34, 45, 52, 78, 119, 122
New York Times, 44, 71
Nixon, Richard: "blue-collar strategy," 43–44; Clean Air Act and, 40; environmental actions of, 5, 39–40, 42; environmental enthusiasm of, 42; EPA formation by, 40; NEPA signed by, 40; populism and, 39; presidential runs of, 37; privatization and, 135n12; "southern strategy," 43; Vietnam War and, 43; Watergate and, 60
Norton, Gale, 109, 111, 113–14
Nye County, Nevada, 79, 140n13

Obama, Barack, 123
Office of Management and Budget (OMB), 69, 72, 142n17
Office of Surface Mining, 66, 72
Oklahoma City bombing, 80–81
Oregon, 4, 14, 32, 67, 100, 103–4
Oregon Natural Resources Council, 78
Outdoor Recreation Resources Review Commission (ORRRC), 32, 34

Pacific Legal Foundation, 50, 98
Pacific Northwest, 75
Patterson, Thomas M., 14
Paul, Rand, 3
Parker v. United States, 46–47
Payne, Ryan, 103, 106, 107
Payton, Jere, 80
Pendley, William, 66, 68
Phillip Burton Wilderness, 57
Pinchot, Gifford, 12–20, 25, 38
Pine Creek Ranch, 94

Point Reyes National Seashore, 57
Potok, Mark, 3, 102
Powder River, 70
Proceedings of the Conference of Governors in Washington, DC (1908), 14
Public Land Law Review Commission, 47
Public Lands Council, 51
Public Works Administration, 22
Pueblo Indians, 102–3

Ray, Dixie Lee, 76
Reagan, Ronald, 5, 39, 56, 59–63, 64–66, 68–73, 108–9
Redwood National Park, 57, 133n22
Reid, Harry, 3
Reno, 78, 81
Republican National Convention (1980), 59
Republican Party, 4; Alaska and, 41; anti-environmentalism, 44, 74, 77; Bundy and, 3; business advocacy of, 29, 43; conservation voters and, 6; conservatism and, 61; Eisenhower and, 31–32; environmentalists and, 44, 74, 112–13; Goldwater faction of, 31–32; internal disagreements of, 16, 19–20, 40; Koch Industries and, 116, 118; land conveyance and, 4, 7, 61, 118, 123; Limbaugh and, 77; "Modern Republicanism" in, 31; Nixon and, 39–40; Reagan and, 59, 61, 64; reducing government and, 64; resource development and, 60, 67
Resource Recovery Act, 40
Revised Statute (RS) 2477, 58, 98
Rhoads, Dean, 51, 62, 83
Ritzheimer, Jon, 104, 105
"roadless rule," 108
Robbers Roost, 34
Robbins, Harvey Frank, 114
Rocky Mountains, 10
Rodrique, Charlotte, 105, 106
Roosevelt, Franklin D., 21, 22–24, 33, 37, 49, 95
Roosevelt, Theodore, 1, 10–12, 14–15, 16–17, 19–20, 21, 104

Rove, Karl, 110
Royal Society (UK), 117
Ruby Valley Treaty of 1863, 84, 89, 91
Ruckelshaus, William, 60, 72
Ryan, Paul, 105

Sabatier, Paul, 66
Sagebrush Rebellion, 5; Andrus on, 53; California and, 53; collapse of, 55, 67, 71; Laxalt on, 62; legislation and, 57; Nevada and, 51, 55; Reagan administration and, 62, 66-67, 71; Watt and, 67, 71
"Sahara Clubbers," 79-80
Salmon River, Idaho, 103, 137n25
San Francisco Chronicle, 17
San Juan County, Utah, 54, 102
Santilli, Pete, 104, 107
Santini, Jim, 56-57, 62, 68-69
Santini-Burton bill, 56, 57
Save the San Francisco Bay Association, 36
Saylor, John, 35
Scaife, Richard Mellon, 64
Science (magazine), 114
Seattle, 78
Shell Oil Company, 70
Shoshone Indians, 89-93
Shovel Brigade, 84, 86, 110
Sierra Club, 30, 31, 33, 34, 72
Sierra Nevada, 10, 30, 56, 83
Silent Spring (Carson), 36
Simpson, Alan K., 68
Smith, Loren A., 97-98
Smith River, 63
South Carolina Ordinance of Nullification, 9
Southern Poverty Law Center, 3, 102
Sparks, William A. J., 10
"State of the Rockies Project," 122
Stegner, Wallace, 34
Stevens, Ted, 49, 50
Stockman, Steve, 77
Storm over Rangelands: Private Rights in Federal Lands (Hage), 95
Swift and Company, 104

Taft, William Howard, 16-19, 20
Tahoe, Lake, 17-18, 56-57, 137n3
Taylor, Edward T., 23
Taylor Grazing Act, 21, 23, 25, 32, 48
Teapot Dome, 20
Teller, H. M., 13
Templeton, Billy, 92
Tenth Amendment, 45, 78, 84
Texas, 4, 69, 116
Thompson, Bruce, 92
Thompson, Mike, 111
Tibbits, Ray, 58
Tixier, Susan, 86, 140n8
Tonopah, Nevada, 94
Tonopah Bombing Range, 130n1
Train, Russell, 42, 60, 69
Trans-Alaska Pipeline, 40-41
Trans-Mississippi Commercial Congress, 13
Trinity River, 63

Udall, Morris, 49, 62, 70
Udall, Stewart, 32-33, 34, 37, 41
Union of Concerned Scientists, 117
Upper Klamath Lake, 110-11
U.S. Army Corps of Engineers, 84
U.S. Chamber of Commerce, 29
U.S. Constitution, 9, 11, 45
U.S. Court of Appeals for the Ninth Circuit, 63, 91-92, 98-100
U.S. Department of Justice, 70, 78, 79, 114
U.S. Department of the Interior, 24-25, 27-28, 32, 40, 45, 47, 109-11, 114
U.S. District Court, District of Columbia, 116
U.S. District Court, District of Nevada, 89
U.S. Division of Grazing, 23-25
U.S. Forest Service: Agriculture Department and, 24-25; creation of, 12, 13; firefighting costs of, 7; grazing and, 23; Hage and, 94-99; Lake Tahoe and, 52; MUSY and, 47, 121; Nixon and, 42, 43; Pinchot and, 12; rangelands and, 47, 120; Sagebrush

Rebellion and, 52, 79, 84–85, 87; timber policy of, 47, 72; wilderness and, 46–47
U.S. General Land Office, 33
U.S. Grazing Service, 23, 25–27
U.S. House of Representatives Interior Committee, 70
U.S. House of Representatives Technology Assessment Office, 70
U.S. Senate, 4, 6, 25, 35, 56, 118
U.S. Supreme Court: equal-footing doctrine and, 69; Dann sisters and, 91; grazing and, 24; Hage case and, 98, 100; interior secretary authority and, 63; ozone and, 116; public domain and, 46; *Stearns v. Minnesota* finding by, 11; wilderness and, 47
U.S. Timber Culture Act of 1873, 10
U.S. Treasury, 91, 97
Utah, 4, 52, 57, 83, 102–3, 122

Veneman, Ann, 111
Vietnam War, 36, 38, 43

Waco siege, 77
Walden, Greg, 105
Wallop, Malcolm, 68
Washington, DC, 4, 43, 104, 122
Washington Post, 34, 65
Washington State, 4, 14, 32, 52, 80
Washington Times, 20
Watergate, 60, 81
Water Pollution Control Act, 42

Watt, James: Andrus and, 63; Congress and, 68, 70–71, 72; economists against, 67–68; environmentalists and, 67–68; good-neighbor policy by, 67, 71; Mountain States Legal Defense Fund and, 50, 65; petition against, 68; public relations and, 65–66, 71–72; Reagan and, 68; regulatory cuts by, 67; Sagebrush Rebellion and, 67, 71
Western Lands bill, 56
Western Wood Products Association, 47
White, Richard, 76
White House Conference on Conservation (1962), 35
Whitman, Christine Todd, 113
Wild and Scenic Rivers Act of 1968, 37, 47
Wilderness Act, 35–36, 46–47
Wilderness Society, 34, 42, 72–73, 87
Wilson, Helen, 86
Wilson, Woodrow, 20
Wise Use movement, 75–76, 78, 80, 104, 110
Wyoming, 4, 14, 52, 68, 70, 122

Yellowstone National Park, 113
Young, Clifton, 51, 53
Young, Don, 118
Yowell, Raymond, 90–91
Yucca Mountain, Nevada, 79
Yurok Tribe, 111

Zahniser, Howard, 34–35, 36

Although born in New York City, Michael J. Makley has spent most of his life in California's Sierra Nevada. He is the author of several books on Western history. *The Infamous King of the Comstock: William Sharon and the Gilded Age in the West* won *Foreword Magazine*'s Silver Award for Biography. Other books include *John Mackay: Silver King in the Gilded Age* and a trilogy about *Lake Tahoe: A Short History of Lake Tahoe; Cave Rock: Climbers, Courts, and a Washoe Indian Sacred Place*, written with his son Matthew; and *Saving Lake Tahoe: An Environmental History of a National Treasure*. He also produced a video, *Cave Rock, the Issue,* used as a source document in a dispute over a Native American cultural site ultimately decided in the Ninth Circuit Court of Appeals. With his wife Randi, he splits time between Alpine County, California, and Denver, Colorado.

www.ingramcontent.com/pod-product-compliance
Lightning Source LLC
Chambersburg PA
CBHW032215230426
43672CB00011B/2564